D1807229

Aniruddha Chandra

Diversity Combining for Digital Signals in Wireless Fading Channels

Aniruddha Chandra

Diversity Combining for Digital Signals in Wireless Fading Channels

Analysis and Simulation of Error Performance

LAP LAMBERT Academic Publishing

Impressum/Imprint (nur für Deutschland/ only for Germany)
Bibliografische Information der Deutschen Nationalbibliothek: Die Deutsche Nationalbibliothek
verzeichnet diese Publikation in der Deutschen Nationalbibliografie; detaillierte bibliografische
Daten sind im Internet über http://dnb.d-nb.de abrufbar.
 Alle in diesem Buch genannten Marken und Produktnamen unterliegen warenzeichen-, marken-
oder patentrechtlichem Schutz bzw. sind Warenzeichen oder eingetragene Warenzeichen der
jeweiligen Inhaber. Die Wiedergabe von Marken, Produktnamen, Gebrauchsnamen,
Handelsnamen, Warenbezeichnungen u.s.w. in diesem Werk berechtigt auch ohne besondere
Kennzeichnung nicht zu der Annahme, dass solche Namen im Sinne der Warenzeichen- und
Markenschutzgesetzgebung als frei zu betrachten wären und daher von jedermann benutzt
werden dürften.

Coverbild: www.ingimage.com

Verlag: LAP LAMBERT Academic Publishing GmbH & Co. KG
Dudweiler Landstr. 99, 66123 Saarbrücken, Deutschland
Telefon +49 681 3720-310, Telefax +49 681 3720-3109
Email: info@lap-publishing.com

Herstellung in Deutschland:
Schaltungsdienst Lange o.H.G., Berlin
Books on Demand GmbH, Norderstedt
Reha GmbH, Saarbrücken
Amazon Distribution GmbH, Leipzig
ISBN: 978-3-8443-9102-2

Imprint (only for USA, GB)
Bibliographic information published by the Deutsche Nationalbibliothek: The Deutsche
Nationalbibliothek lists this publication in the Deutsche Nationalbibliografie; detailed
bibliographic data are available in the Internet at http://dnb.d-nb.de.
 Any brand names and product names mentioned in this book are subject to trademark, brand
or patent protection and are trademarks or registered trademarks of their respective holders.
The use of brand names, product names, common names, trade names, product descriptions
etc. even without a particular marking in this works is in no way to be construed to mean that
such names may be regarded as unrestricted in respect of trademark and brand protection
legislation and could thus be used by anyone.

Cover image: www.ingimage.com

Publisher: LAP LAMBERT Academic Publishing GmbH & Co. KG
Dudweiler Landstr. 99, 66123 Saarbrücken, Germany
Phone +49 681 3720-310, Fax +49 681 3720-3109
Email: info@lap-publishing.com

Printed in the U.S.A.
Printed in the U.K. by (see last page)
ISBN: 978-3-8443-9102-2

Preface

Fading and its mitigation played a pivotal role in radio communication since its inception. Fading results from the multipath environment of the radio propagation. Since the channel for radio communication is time-varying and random, the modelling of the fading process should therefore be of statistical nature. Several distributions have been derived for the fading model, of which Rayleigh and Rician distribution have been extensively used in the early phases of the radio communication. However, the complex environments of mobile communication demanded a relook of these models, which further led to more and more sophisticated models of fading. Nevertheless, none of the models can describe the fading environment of the radio channel completely and accordingly the different models remain in use depending on the situation, or the environment.

The advent of digital communication and the requirement of robust control on power and bandwidth necessitated use of large number of competing modulation schemes for the digital signals, each having its own advantages and disadvantages. The existence of such a large number of modulation techniques requires that the phenomenon of fading be studied for each and every modulation scheme. Studying some of these are simple and can easily be converted into an analytical expression, whereas dealing with some of them turned out to be strenuous with no analytic expressions.

The diversity combining has traditionally been used as a mitigation to the derogatory effect of fading. Several diversity combining methods have evolved over time with varying degrees of robustness against fading. Implementation of these techniques, to a large extent, is determined on cost and complexity as well as on the type of applications for which they are to be used.

Intensive research on fading and diversity combining mechanisms are already reported in the literature. However, new scope in this field are emerging with the new applications in the wireless communications, and the field remains fertile for further studies. The author, in this book, investigates effect of fading on different modulation techniques hitherto untrodden and find improvements using the combining methods. The main thrust is to find out analytical expressions in closed-form where possible and /or techniques to establish simulation models closely resembling the actual communication environment. To establish the efficacy of analytical models, the closed-form analytical results have also been compared with respect to the simulation models and vice-versa. The study involves the performance of digital modulated signals under different fading models such as Rayleigh, Rician, Nakagami, and Hoyt fading followed by the improvement due to diversity combining schemes such as maximal ratio combining, equal gain combining, selection combining, switch and stay combining, and switch and examine combining. The results are consistent with the previous studies but are more accurate and generated with simple models.

This book is largely based on my doctoral thesis work and would never be possible without the full support and constant encouragement from my thesis supervisor, Dr. Chayanika Bose, who serves as a Reader in Jadavpur University, Kolkata, India. During my thesis tenure, I had the great fortune and honour to discuss with Dr. Manas Kr. Bose of Ansaldo–STS on several problems, and he kept me updated about the recent developments. Present and past faculty members of my current workplace, NIT Durgapur, have helped me a lot by providing a stimulating environment for research. I would also like to thank my students (as well as co-author of various research papers) for their assistance. Last, but certainly not the least, I would like to acknowledge the commitment, sacrifice and support of my parents and my wife.

NIT Durgapur   Aniruddha Chandra
May 2011

Contents

Chapter 3: Fading, Diversity and Combining 52

List of Figures

List of Tables

List of Acronyms

AP	Access point
AWGN	Additive white Gaussian noise
BEP	Bit error probability
BPSK	Coherent binary phase shift keying
BFSK	Coherent binary frequency shift keying
BS	Base station
CDF	Cumulative distribution function
CEP	Conditional error probability
CHF	Characteristic function
CSI	Channel state information
DBPSK	Differentially coded and coherently demodulated BPSK
DPSK	Differential phase shift keying
EGC	Equal gain combining
IID	Independent and identically distributed
ISI	Inter symbol interference
LC	Linear combining
LOS	Line-of-sight
MDPSK	M-ary differential phase shift keying
MFSK	Coherent M-ary frequency shift keying
MGF	Moment generating function
MIMO	Multiple input multiple output
MPSK	M-ary phase shift keying
MQAM	M-ary quadrature amplitude modulation
MRC	Maximal ratio combining
MSK	Minimum shift keying
ncBFSK	Non-coherent binary frequency shift keying
ncMFSK	Non-coherent M-ary frequency shift keying
PDF	Probability density function

PSD	Power spectral density
QoS	Quality of Service
RV	Random variable
SC	Selection combining
SEC	Switch-and-examine combining
SEP	Symbol error probability
SNR	Signal to noise ratio
SSC	Switch-and-stay combining
SWC	Switched combining
WLAN	Wireless local area network

List of Symbols

$B_z(\cdot,\cdot)$	Incomplete Beta function
$\mathrm{erf}(\cdot)$	Error function
$\mathrm{erfc}(\cdot)$	Complementary error function
E_s	Energy per symbol
E_b	Energy per bit
$f(\cdot)$	Probability density function
$F(\cdot)$	Cumulative distribution function
$_1F_1(\cdot;\cdot;\cdot)$	Confluent hypergeometric function
$_2F_1(\cdot,\cdot;\cdot;\cdot)$	Gauss hypergeometric function
$F_1(\cdot;\cdot,\cdot;\cdot;\cdot,\cdot)$	Appell's hypergeometric function of first kind
$F_2(\cdot;\cdot,\cdot;\cdot,\cdot;\cdot,\cdot)$	Appell's hypergeometric function of second kind
$F_A(\cdot;\cdots;\cdots;\cdots)$	Lauricella's multiple hypergeometric function
$_pF_q\left(\{\alpha_1,\cdots,\alpha_p\};\{\beta_1,\cdots,\beta_q\};\cdot\right)$	Generalized hypergeometric function
$H_n(\cdot)$	Hermite polynomial
$I_\nu(\cdot)$	Modified Bessel function of the first kind
$In_z(\cdot,\cdot)$	Ratio of Beta functions
K	Rician fading parameter
$K_\nu(\cdot)$	Modified Bessel function of the second kind
L	Number of diversity branches
$L_n^\alpha(\cdot)$	Laguerre polynomial
m	Nakagami fading parameter
M	Modulation order
$M_\gamma(\cdot)$	Moment generating function
$M_{\mu,\nu}(\cdot)$	Whittaker's function of the first kind

P_e	Symbol error probability
$P_{e,b}$	Bit error probability
P_o	Outage probability
$P_\nu^\mu(\cdot)$	Associated Legendre function of the first kind
q	Hoyt fading parameter
$Q(\cdot)$	Gaussian Q function
$Q(\cdot,\cdot)$	First order Marcum's Q function
$Q_m(\cdot,\cdot)$	mth-order Marcum's Q function
$Q_\nu^\mu(\cdot)$	Associated Legendre function of the second kind
T_s	Duration of a symbol
T_b	Duration of a bit
$U(\cdot;\cdot;\cdot)$	Tricomi's confluent hypergeometric function
$W_{\mu,\nu}(\cdot)$	Whittaker's function of the second kind
γ	SNR per symbol in fading channel
$\bar{\gamma}$	Average SNR per symbol in fading channel
γ_b	SNR per bit in fading channel
$\bar{\gamma}_b$	Average SNR per bit in fading channel
$\gamma(\cdot,\cdot)$	Incomplete Gamma function
$\Gamma(\cdot)$	Gamma function
$\Gamma(\cdot,\cdot)$	Complementary incomplete Gamma function
η	SNR per symbol in AWGN channel
η_b	SNR per bit in AWGN channel

Chapter 1

Introduction

Wireless channels are rich in random scatterers which lead to multiple reception of the same transmitted signal with different delays. Therefore, the received signal is, in general, a sum of the components reaching the receiver from different paths. This multipath propagation causes unwanted variation of the received signal; popularly known as *fading*. Unlike wired channel that is stationary and predictable, for wireless channels the strength of the received signal may fluctuate rapidly with respect to time and the position of the receiver relative to the transmitter. Further, the time-varying nature of the channel, presence of Doppler effect due to relative motion, and frequency selectivity due to delay spread render any sort of wireless communication quite difficult in terms of coverage, capacity, and quality of service (QoS).

The simplest technique to maintain acceptable grade of service and channel capacity is to increase the *transmitted power*. Wireless mobile terminals are generally small, lightweight, battery-driven, low-power devices which restrict their capability of increasing transmission power; especially in the up link. Even if higher power level requirement is fulfilled, the overall spectrum efficiency (number of transmitted bits per bandwidth per unit area of coverage) is reduced because the distance between frequency reuse transmitters must be greater to maintain tolerable co-channel interference (CCI). The adjacent channel interference (ACI) felt will also go up. Moreover, fading sometimes generates irreducible error rates that cannot be overcome simply by increasing the transmitted power.

One might think of sacrificing *bandwidth* (BW), the other major resource in a communication system, to mitigate fading. Error control coding schemes may be used for trading off reliability at the cost of bandwidth via controlled redundancy i.e., sending additional information about the transmitted signal. Unfortunately bandwidth is another scarce resource for wireless applications due to stringent regulations. The radio frequency spectrum is already crowded with the varied and abundant use of wireless communications. Data traffic in wireless networks had surpassed the voice traffic long ago and new high-speed data and multimedia services are being introduced at a rapid pace which rather

1

demand more spectrally efficient system design. In addition, fading often leads to burst errors, while error correction coding schemes are best suited for discrete errors.

On the other hand, one resource that is growing at a very rapid rate is the *processing power* of the communication nodes. Given these circumstances, there has been considerable research effort in recent years aimed at development of novel signal transmission techniques and advanced receiver signal processing methods. These sophisticated spectral and power efficient fade mitigation techniques improve radio link performance significantly. In particular, *diversity combining* techniques have great potential to improve the performance of wireless systems without an increase in the transmitted power and bandwidth. This is one of the reasons for a new surge of interest in the studies of fading and mitigation techniques, and motivated this author to investigate the existing techniques and tools of diversity combining with an aim to improve them further.

1.1   Modelling and Methods

It is important to assess the efficacy of different receiver design options so as to determine the most appropriate choice of combining scheme and modulation method regarding performance, complexity and implementation constraints. To undertake this task, one may resort to either simulation modelling (usually Monte Carlo type) or, consider developing an analytical framework. The analytical approach has three obvious advantages over the Monte Carlo simulation methodology: first, it facilitates rapid computation of the system performance; second, it provides insight as to how different design parameters affect the overall system performance; and third, it becomes easier to optimize the design parameters. On the other hand, simulation models may provide valuable results in case the theoretical analysis becomes too difficult and complex. The analytical solutions are ideal for use in the first-step of the design process of a communication system, because the framework allows one to quickly narrow down the viable design options to only a few, before an extensive computer simulation effort is undertaken. In the current text, the plan is to develop both types of models simultaneously as far as possible.

The study is limited to pre-detection diversity combining. Although post-detection systems provide better performance, they require multiple RF chains associated with multiple antennas, which are costly in terms of size, power, and hardware. Also, more recent work has focused on investigating how techniques such as transmit diversity and space time codes can be applied to maximize diversity gain. The disadvantage of spacetime codes is the high decoding complexity, which grows exponentially as a function of both the required capacity and diversity order, and has not been considered in this book.

1.2 Contributions of the Book

Fading had been a subject of intense research since the early days of wireless communications. There has been a widespread belief that the problem of fading and the corresponding mitigation techniques have been exhaustively studied, leaving a little room for original contributions. However, the ever expanding horizon of wireless and mobile communication requires that the problem of fading and its mitigation be studied more meticulously to assimilate these new concepts for realization of reliable communication over the congested radio spectrum. With a view to the above, this study focuses on refining the existing mathematical tools and modelling methods to the latest mobile communication requirements. Specifically the contributions of this book are the following:

1. Systematic study of error performance calculation under a unified analytical framework

    (a) Unified analysis of binary modulations with different diversity combining methods in Rayleigh fading channel (Chapter 4).

    (b) Unified analysis of coherent modulations with maximal ratio combining (MRC) in Rician fading channel (Chapter 5).

2. Refining the existing analytical approaches, proving the equivalence of different formulas, and bridging the gap

    (a) Improved approximations for symbol error probability (SEP) of coherent M-ary frequency shift keying (MFSK) with MRC diversity over different fading channels (Chapter 4-7).

    (b) Series solutions for bit error probability (BEP) of $\pi/4$ shifted differential quadrature phase shift keying ($\pi/4$-DQPSK) with MRC diversity over different fading channels (Chapter 4-7).

    (c) Closed analytical form for BEP expressions of $\pi/4$-DQPSK with selection combining (SC) in Nakagami fading channel (Chapter 6).

    (d) Derivation of SEP expressions for non-coherent M-ary frequency shift keying (ncMFSK) in Rician fading channel (Chapter 5) and in Hoyt fading channel (Chapter 7).

3. Finding new analytical expressions for

    (a) BEP of binary modulations with switch and examine combining (SEC) in double Rayleigh fading channel (Chapter 4).

    (b) Cumulative distribution function (CDF) of the instantaneous signal to noise ratio (SNR) for Hoyt fading channel (Chapter 7).

3

(c) Symbol error probability of M-ary modulation schemes with SEC in Hoyt fading channel (Chapter 7).

4. Modelling and simulation of different modulation processes, fading channels, and diversity combining methodologies (Chapter 8).

1.3 Organization of the Book

This book examines error performance of digital modulated signals over wireless fading channels. A thorough description of these modulation schemes and their performance in absence of fading are prerequisites for developing the error expressions over fading channels, which are provided in Chapter 2. The chapter begins with the basic transmission model and different representations of modulated digital signals. Thereafter, constellation diagrams, receiver architectures, and error probability calculations are discussed for each and every modulation technique considered in the book. The chapter concludes with a brief comparison of the modulation schemes on the basis of error probability and bandwidth-SNR tradeoff achieved through the modulation processes.

Chapter 3 describes the mobile wireless channel, signal degradations due to different fading mechanisms, and various diversity techniques that can be used to overcome the effects of fading. The impulse response of a multipath time-varying channel is developed, and it is shown that depending on the propagation environment, the attenuation profile of the generic channel model obeys different statistical distributions (e.g., Rayleigh, Rician etc.). Various means of achieving diversity is discussed next, highlighting receiver space diversity, and pointing out its merits over others. The performance of any diversity system depends on the combining technique used to merge the signals received from the disparate diversity branches, and there exists a lot of such techniques. In Chapter 3, four basic combining methods and their compatibility with coherent and non-coherent modulations are discussed.

Rayleigh distribution had been used for more than half a century for modelling wireless fading channel with limited mobility, and remains the most acceptable model till date. Recently, for modelling the highly mobile fading channels, double or cascaded Rayleigh distribution has been introduced. In Chapter 4, a unified approach is presented for calculation of error probability of binary modulations over both single and double Rayleigh fading channels with different diversity combining techniques. Error performance for other M-ary modulations namely, $\pi/4$-DQPSK, coherent MFSK, and M-ary quadrature amplitude modulation (MQAM), in presence of diversity, are also derived.

Chapter 5, on the other hand, is concerned with the Rician fading channel, where BEP of binary and $\pi/4$-DQPSK modulation is presented in a simpler form. In addition, SEP of both coherent and non-coherent MFSK is found in presence of suitable diversity com-

4

bining (MRC and SC respectively). A unified approach for evaluating error performance of several coherent multichannel modulation schemes is proposed next, which includes MQAM as a special case.

In Chapter 6, closed-form BEP expressions for $\pi/4$-DQPSK with SC and MRC over Nakagami fading channel are derived. Performance of coherent MFSK with MRC is also included. Nakagami fading model had been studied exhaustively in the past due to its wide capability of modelling a wide class of fading environments and the analytical tractability of the fading distribution. In fact, calculations for all other modulation schemes are either available in the open literature or can be easily derived, and therefore not included in this book.

On the contrary, Chapter 7 deals with a relatively less studied fading model, Hoyt distribution. At first, a closed-form expression of CDF of the instantaneous SNR in Hoyt channel is derived. This CDF and associated formulas are then used to find out the error probability of different modulation schemes, $\pi/4$-DQPSK, MFSK, and ncMFSK. The switched diversity case is covered in a bit more detail, and some additional performance parameters like, outage probability and average SNR with switched diversity, are provided.

Modelling and simulation of different modulation processes, fading channels, and diversity combining methodologies were necessary to validate the analytical results presented in the book. The final chapter, Chapter 8, gives a brief account of the same.

Chapter 2

Performance of Digital Signals with AWGN

Wireless systems developed in the 80's were essentially analog and most of them used frequency modulation (FM) for carrying information across the channel. For example advanced mobile phone system (AMPS), a first generation (1G) wireless standard popular in north America, used to operate in the 800 MHz cellular FM band. To meet ever-increasing demand for voice and data communication, from second generation (2G) onwards digital systems started to replace old analog devices. Digital transmission offers better noise immunity, is more resistant to channel impairments like interference or fading, facilitates use of regenerative repeaters, provides with a wider choice of multiple access/multiplexing schemes, and can realize better security/privacy through encryption. These features added with the power of flexible signal processing options and advances in digital hardware over past few decades make digital platform a natural choice for wireless transceivers being built or proposed. Furthermore, digital modulations are also suitable for efficient realization of trade-offs between fundamental communication resources (transmitter power and channel bandwidth) and quality of service metrics (e.g. probability of error, P_e) over a power-limited, bandwidth-limited channel like wireless medium.

Digital modulation is a process for converting discrete-time symbols that belong to a finite alphabet into continuous-time signals suitable for transmission through the physical channel between modulator and demodulator [1]. The mapping is, in general, one-to-one. Although not mandatory for wired channels, in wireless channels modulation essentially involves frequency up-conversion. The demand for this frequency translation arises for realizing a suitable antenna dimension. For both wired and wireless channels, the signal, before reaching the receiver, gets distorted during propagation through channel due to inherent noise present in the system. In absence of any such noise, a demodulator block at receiver end is able to fully reverse the action of a modulator without any error. However, for practical channels demodulation refers to frequency down-conversion and inverse mapping to extract the original symbols from a noisy version of the transmitted signal.

If the output of a modulator during a particular signalling interval depends on one or more previously transmitted waveforms, the modulator is said to have *memory*. On the other hand, when the mapping from the symbol to waveform is performed without any dependency on previously transmitted waveforms, the modulator is called *memoryless*. In addition to classifying the modulator as either memoryless or having memory, we may classify it as either *linear* or, *non-linear* [2]. Linearity of a modulation method requires that the principle of superposition applies in the mapping of the digital sequence into successive waveforms. Amplitude and phase modulation falls under this category. Linear modulations have better spectral efficiency [3] but require expensive and less power efficient linear amplifiers. In non-linear modulation, the superposition principle does not apply to signals transmitted in successive time intervals. Since frequency modulation typically has a constant signal envelope and is generated using nonlinear method, it belongs to the class of constant envelope modulation or non-linear modulation. Non-linear modulations are less susceptible to channel impairments (fading and interference), but non-linear processing leads to spectral broadening [4]. Finally, digital modulations may also be sub-divided into two groups - *coherent* and *non-coherent*, depending on the demodulator structure. When the receiver exploits knowledge of the carrier wave's phase reference to detect the signals, the process is called coherent detection. In ideal coherent detection, prototypes of the possible arriving signals are available at receiver. During detection, the receiver multiplies and integrates (correlates) the incoming signal with each of its prototype replicas. Non-coherent modulation refers to systems designed to operate with no knowledge of phase. As phase estimation processing is not required, system complexity is reduced, but increased P_e is the trade-off.

Throughout this book our discussion is limited to performance evaluation of memoryless modulation methods, but we address both linear and non-linear modulations as well as coherent and non-coherent detection. In Section 2.1 we describe the transmission model followed by different representations of digitally modulated signals in Section 2.2. Section 2.3 gives detailed error probability expressions for various binary and M-ary modulation schemes. Finally in Section 2.4 a comparison of different modulation schemes is presented. The section summarizes the important aspects discussed in the previous sections and concludes the chapter.

2.1 Transmission Model

2.1.1 Modulator

In the transmission of digital information over a communication channel, the modulator is the interface device that maps the digital information into analog waveforms that match the characteristics of the channel. The mapping is generally performed by taking blocks of $n = \log_2 M$ binary digits at a time from the information sequence $\{a_i\}$ and selecting one of $M = 2^n$ deterministic, finite energy waveforms $s_i(t); i \in \{0, 1, \ldots, M - 1\}$ for transmission over the channel.

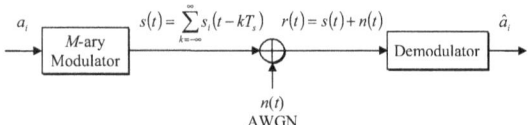

Figure 2.1: Generic transmission model for AWGN channel.

A model for a single-sender, single-receiver digital communication system is depicted in Fig. 2.1 where, at every T_s seconds, in response to an input a_i from a set of M alphabets, the modulator produces a signal $s_i(t)$. Inverse of the *symbol duration* T_s, i.e. $R_s = 1/T_s$ is known as the signalling rate in symbols per second or *baud rate*. Usually, $M(= 2^n)$ is an integer power of 2, so we can think of each symbol as attempting to convey n bits per symbol. Hence the bit duration $T_b = T_s/\log_2 M = T_s/n$ and the bit rate would be $R_b = nR_s$ bits per second (bps).

Further, we would assume that signals are amplitude normalized so that at the transmitter the expected energy expended per symbol is E_s joules, where we define electrical energy relative to a 1Ω impedance. Thus average energy per symbol:

$$E_s \triangleq E\left\{ \int_0^{T_s} s_i^2(t)dt \right\} = \frac{1}{M} \sum_{i=0}^{M-1} \int_0^{T_s} s_i^2(t)dt \tag{2.1}$$

where the expectation $E\{\cdot\}$ is taken with respect to the signal index i, and we have assumed equiprobable signal selection, i.e. the *a priori* probabilities of the signals $\Pr[s_i] = 1/M$. Quite naturally average energy per bit $E_b = E_s/\log_2 M = E_s/n$, would be the average symbol energy distributed over all the constituent bits.

Lastly, the passband output of a memoryless digital modulator may be expressed as

$$s(t) = \sum_{k=-\infty}^{\infty} s_i(t - kT_s) \quad ; kT_s \leq t < (k+1)T_s \tag{2.2}$$

representing a superposition of individual signal waveforms transmitted during each signalling interval.

2.1.2 Channel

The modulated signal $s(t)$ passes through a non-ideal channel and reaches the receiver. As per the model shown in Fig. 2.1, the received signal at demodulator input

$$r(t) = s(t) + n(t) \tag{2.3}$$

is perturbed by $n(t)$, a Gaussian real bandpass noise having two-sided power spectral density (PSD)

$$S_n(f) = \frac{N_0}{2} \quad ; -\infty < f < \infty \tag{2.4}$$

A flat PSD over the whole spectrum renders the noise to be *white*. The autocorrelation function of the noise process $n(t)$ is given by the Fourier inverse of the PSD

$$\mathcal{F}^{-1}\{S_n(f)\} = R_n(t_1, t_2) = \frac{N_0}{2}\delta(\tau) \tag{2.5}$$

where $\delta(\cdot)$ is the Dirac's delta function and $\tau = t_1 - t_2$. Eq. (2.5) denotes that the noise samples at different time instants t_1 and t_2 are dissimilar. Also samples from the stationary noise process $n(t)$ forms a random variable (RV) $n \sim \mathcal{N}(0, N_0/2)$, following a *Gaussian* distribution with zero mean and variance $N_0/2$. The corresponding probability density function (PDF) is given by

$$f_n(x) = \frac{1}{\sqrt{\pi N_0}} \exp\left(-\frac{x^2}{N_0}\right) \quad ; \quad \infty < x < \infty \tag{2.6}$$

Furthermore, the noise process is independent of and *additive* to the signal. Combining all these aspects, the noise model is commonly referred to as *additive white Gaussian noise* (AWGN) and is the archetypal communication noise model for radio and optical frequency systems.

Origin of AWGN may be owed to mainly three phenomena [1]: thermal noise in electrical circuits, shot noise processes developed in electronic or photonic devices, and electromagnetic radiation from the Earth, Sun and other cosmic sources of radiation. However, the noise model discussed here does not account for non-Gaussian noise, which may be present due to interference from other signals, electrostatic discharges in the atmosphere, or electrical machinery. The non-white noise is excluded at the outset. Usually the additive noise processes have power spectra that are nearly constant over the range of frequencies larger than that occupied by the signal.

There are certain other assumptions incorporated when we assume an AWGN channel model only. First, we assume that the channel has no bandlimiting effect that may cause time-dispersion or, inter symbol interference (ISI). In other words, the channel can support infinite bandwidth signal. Second, the channel's amplitude response is flat over the whole frequency range and its phase response is linear. This eradicates any possibility of attenuation or, phase distortion. However, this condition is supplemented in later chapters, to incorporate the effect of channel fading. Third, throughout this book, uncoded transmission through channel has been assumed, so the error rates derived do not consider the effect of compression or, controlled redundancy to control errors. Lastly, the receiver is modeled in a manner such that it has complete knowledge of the channel state information (CSI), i.e. the functional forms of transmitted signals $\{s_i(t)\}_{i=0}^{M-1}$, channel gain, time delay and phase information of the signal at demodulator input and so on. This implies that the signals are received with perfect phase, frequency and timing synchronization.

To simplify modeling, the frequency up-conversion at modulator and down-conversion at demodulator may be combined with the real channel to yield an equivalent complex baseband channel model. The transmission through such a channel in the baseband can be expressed as

$$\tilde{r}(t) = \tilde{s}(t) + \tilde{n}(t) \tag{2.7}$$

where $\tilde{n}(t)$ denotes a zero-mean circularly symmetric complex Gaussian process and $\tilde{s}(t), \tilde{r}(t)$ represent baseband equivalent of the transmitted and received signals respectively. A detailed discussion on baseband representation may be found in Section 2.2.1.

2.1.3 Demodulator

The task of the demodulator is to process the received signal $r(t)$ in order to produce an estimate $\hat{a}_i$ that is optimum in the sense that the probability of error,

$$P_e \overset{\triangle}{=} \Pr\left[\hat{a}_i \neq a_i\right] \tag{2.8}$$

is minimum [5].

Design of an optimum demodulator is, in general, a non-trivial task. In order to minimize error probaility, the optimum demodulator sets $\hat{a}_i = a_i$ whenever

$$\Pr[s_i|r] \geq \Pr[s_j|r] \quad : j \in \{0, 1, \ldots, M-1\}, j \neq i \tag{2.9}$$

As $\Pr[s_j|r]$ denotes the probability of signal $s_j(t)$ computed after observing the channel output $r(t)$, it is termed as the *a posteriori* probability of the corresponding signal and such a demodulator is called *maximum a posteriori* (MAP) demodulator. Using Bayes' theorem

$$\Pr[s_j|r] = \frac{\Pr[r|s_j]\Pr[s_j]}{\Pr[r]} \tag{2.10}$$

and noting that $\Pr[r]$ is independent of index j, we may conclude from (2.9) and (2.10) that after observing $r(t)$ the MAP demodulator sets $\hat{a}_i = a_i$ whenever

$$\Pr[s_i]\Pr[r|s_i] \geq \Pr[s_j]\Pr[r|s_j] \tag{2.11}$$

On the other hand, a demodulator that sets $\hat{a}_i = a_i$ whenever

$$\Pr[r|s_i] \geq \Pr[r|s_j] \quad ; j \in \{0,1,\ldots,M-1\}, j \neq i \tag{2.12}$$

regardless of the a priori signal probabilities is called *maximum likelihood* (ML) demodulator. Such a demodulator is often used when the a priori probabilities are not known. For equiprobable signal selection, $\Pr[r|s_i] = 1/M$ and the ML decision rule becomes identical to the MAP decision rule. Also under this condition the ML demodulator minimizes the probability of decision error. The prior signal probabilities will be equal when the source coding is efficient.

Optimal ML demodulators that utilize the phase information of the incoming signal belong to the coherent category. A coherent demodulator may be realized [1] with a bank of M correlators where the jth correlator produces an output r_j proportional to the likeliness of $r(t)$ with $s_j(t)$,

$$r_j = \int_{kT_s}^{(k+1)T_s} r(t)s_j(t)dt \quad ; j \in \{0,1,\ldots,M-1\} \tag{2.13}$$

At sampling times $t = kT_s$ the coherent demodulator compares the M decision variables $\{r_j\}_{j=0}^{M-1}$ and decides on the hypothesis corresponding to the largest of them. Fig. 2.2 depicts a functional realization of optimal demodulator with correlator receiver.

For non-coherent demodulators, correlators are replaced by matched filters. In this case the decision variables $\{r_j\}_{j=0}^{M-1}$ are generated with a bank of linear filters. The impulse response of the jth filter $h_j(t)$ is matched to $s_j(t)$. In other words, $h_j(t)$ is a folded and time-shifted version of $s_j(t)$

$$h_j(t) = s_j(T_s - t) \quad ; 0 \leq t \leq T_s \tag{2.14}$$

The matched filter version of the optimal demodulator is shown in Fig. 2.3.

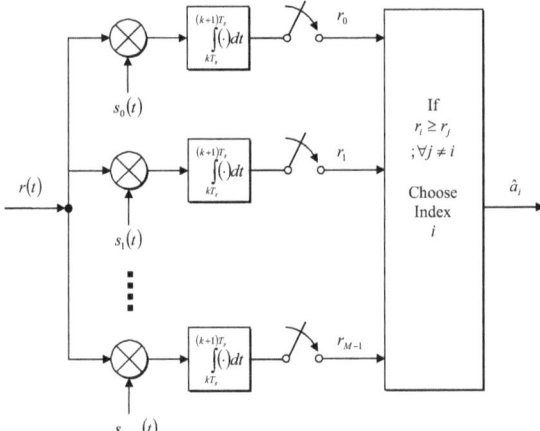

Figure 2.2: Optimal demodulator in correlator form for equiprobable, equal energy signals.

2.2 Representation of Digital Signals

For the general M-ary case transmitted signal $s(t)$, throughout a symbol duration T_s, belongs to a finite set $s_i(t) \in \{s_0(t), s_1(t), \dots s_{M-1}(t)\}$ of waveforms having finite energy. These waveforms may differ in amplitude, phase, frequency, or a combination of these parameters determined by the type of modulation process. A general representation of $s_i(t)$ is

$$s_i(t) = A_i \cos\left[2\pi(f_c + f_i)t + \phi_i\right] \quad ; 0 \leq t \leq T_s \tag{2.15}$$

where $i \in \{0, 1, \dots, M-1\}$.

Amplitude of the ith waveform A_i is related to the corresponding signal energy as

$$A_i = \sqrt{\frac{2E_i}{T_s}} \quad \text{where} \quad E_i = \int_0^{T_s} s_i^2(t)dt \tag{2.16}$$

In case of phase or frequency modulation, signals carry equal energy ($E_i = E_s$) and the amplitude may be written as

$$A_i = \sqrt{\frac{2E_s}{T_s}} \tag{2.17}$$

The phase term ϕ_i in (2.15) denotes the phase associated with the ith signal waveform

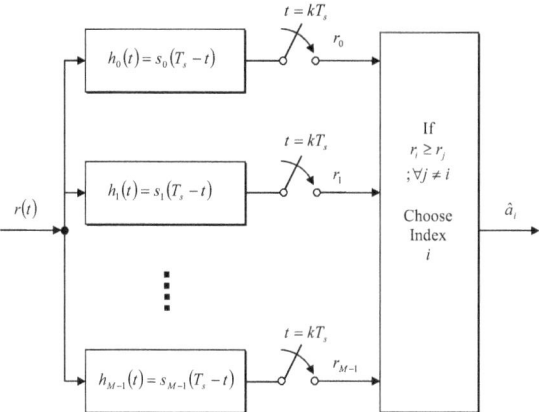

Figure 2.3: Optimal demodulator in matched filter form for equiprobable, equal energy signals.

whereas the frequency $f_c + f_i$ is set in the vicinity of some carrier frequency (f_c). Generally the carrier is a high frequency continuous wave (CW) suitable for transmission with the frequency f_c being an integral multiple of $1/T_s$.

2.2.1 Equivalent Baseband Representation

Section 2.1 describes how a baseband message a_i is mapped to an analog CW signal $s_i(t)$ through digital modulation and then fed to the wireless channel. The resultant signals traveling through the channel are narrowband signals, having a frequency spectra that occupies a narrow bandwidth of $2W$ centered around carrier frequency f_c where, $W \ll f_c$. With the help of these notations we may formally define a bandpass signal $s_i(t)$ whose frequency domain representation $S_i(f)$ is non-zero for frequencies in a small neighborhood of some high frequency f_c, i.e.

$$S_i(f) \approx 0 \quad \text{for } |f - f_c| > W \text{ where } W \ll f_c \tag{2.18}$$

If $s_i(t)$ is a real bandpass signal of finite energy it may be expressed as [3]

$$s_i(t) = \Re\{\tilde{s}_i(t) \exp(j2\pi f_c t)\} \tag{2.19}$$

where $\tilde{s}_i(t)$ is known as the *equivalent baseband* signal or, *complex envelope* of $s_i(t)$. In general $\tilde{s}_i(t)$ is complex

$$\tilde{s}_i(t) = \tilde{s}_{i,I}(t) + j\tilde{s}_{i,Q}(t) \tag{2.20}$$

having an in-phase part, $\tilde{s}_{i,I}(t)$ and a quadrature part, $\tilde{s}_{i,Q}(t)$ respectively. By expanding the exponential term in (2.19) with Euler's relation $\exp(j\theta) = \cos\theta + j\sin\theta$ and substituting the definition of $\tilde{s}_i(t)$ from (2.20), the bandpass waveform can also be expressed in the quadrature form

$$s_i(t) = \tilde{s}_{i,I}(t)\cos(2\pi f_c t) - \tilde{s}_{i,Q}(t)\sin(2\pi f_c t) \tag{2.21}$$

The quadrature designation derives from the fact that in (2.21) second term has a carrier which is $\pi/2$ shifted with respect to the original carrier.

Using the relation $\cos(A+B) = \cos A\cos B - \sin A\sin B$, the general modulated signal given by (2.15) may be represented as

$$s_i(t) = A_i[\cos(2\pi f_c t)\cos(2\pi f_i t + \phi_i) - \sin(2\pi f_c t)\sin(2\pi f_i t + \phi_i)] \tag{2.22}$$

Noting the similarity between (2.21) and (2.22) it is easy to verify that the corresponding in-phase and quadrature components of the complex envelope are given by

$$\tilde{s}_{i,I}(t) = A_i\cos(2\pi f_i t + \phi_i) \tag{2.23a}$$
$$\tilde{s}_{i,Q}(t) = A_i\sin(2\pi f_i t + \phi_i) \tag{2.23b}$$

and the equivalent baseband representation of modulated signal in (2.15) is

$$\tilde{s}_i(t) = A_i\exp[j(2\pi f_i t + \phi_i)] \tag{2.24}$$

It is well known that all demodulators are equipped with a bandpass filter (BPF) of bandwidth $2W$ that is placed at the receiver front end. The BPF only allows signal frequencies in the range $f_c - W \leq f \leq f_c + W$. Thus, although the modeled noise is having an infinite spectrum, the actual noise signal $n(t)$ deteriorating signal reception is also a bandlimited signal. Applying similar arguments as presented above, we may have the following decomposition of a passband noise signal:

$$n(t) = \tilde{n}_I(t)\cos(2\pi f_c t) - \tilde{n}_Q(t)\sin(2\pi f_c t) \tag{2.25}$$

As $n(t)$ is a zero-mean Gaussian process with variance $N_0/2$, the in-phase term $\tilde{n}_I(t)$ and quadrature term $\tilde{n}_Q(t)$ also represent zero-mean Gaussian processes but with variance N_0.

The equivalent baseband noise

$$\tilde{n}(t) = \tilde{n}_I(t) + j\tilde{n}_Q(t) \tag{2.26}$$

is therefore complex circular with PSD N_0 per quadrature.

2.2.2 Signal Space Representation

The signal set $\{s_i(t)\}_{i=0}^{M-1}$ of M finite-energy, finite-duration waveforms can be expressed as linear combinations of N orthonormal basis functions $\{\varphi_0(t), \varphi_1(t), \ldots, \varphi_{N-1}(t)\}$ as,

$$s_i(t) = \sum_{j=0}^{N-1} s_{ij}\varphi_j(t) \quad ; i \in \{0, 1, \ldots, M-1\} \tag{2.27}$$

where $N \leq M$ and the equality holds when the set $\{s_i(t)\}_{i=0}^{M-1}$ is linearly independent. The coefficients s_{ij} of the expansion may be obtained from the following relation

$$s_{ij} = \int_0^{T_s} s_i(t)\varphi_j(t)dt \tag{2.28}$$

The basis functions $\{\varphi_0(t), \varphi_1(t), \ldots, \varphi_{N-1}(t)\}$ are orthonormal which means that

$$\int_0^{T_s} \varphi_i(t)\varphi_j(t)dt = \delta_{ij} \tag{2.29}$$

where δ_{ij} is the Kronecker's delta function defined as,

$$\delta_{ij} = \begin{cases} 1 & , i = j \\ 0 & ; i \neq j \end{cases} \tag{2.30}$$

Exact functional form of the set $\{\varphi_j(t)\}_{j=0}^{N-1}$ depends on the original set of signal waveforms $\{s_i(t)\}_{i=0}^{M-1}$. An appropriate set of basis functions may be constructed through Gram-Schmidt orthogonalization procedure.

2.2.3 Vector Space Representation

From the discussion in Section 2.2.2 it is apparent that each signal $s_i(t)$ is completely characterized by the signal vector

$$\mathbf{s_i} = [s_{i0} \quad s_{i1} \quad \cdots \quad s_{iN-1}]^T \in \mathcal{R}^N \quad ; i = 0, 1, \ldots, M-1 \tag{2.31}$$

of its coefficients s_{ij}. The set of signal vectors $\{\mathbf{s_i}\}_{i=0}^{M-1}$ defines a corresponding set of M points known as *signal constellation* in an N-dimensional Euclidean space called *signal space*. Further, the N mutually perpendicular axes of the signal space in $\mathcal{R}^N$ are the orthonormal basis functions $\{\varphi_j(t)\}_{j=0}^{N-1}$. The vector $\mathbf{s_i} \in \mathcal{R}^N$ denotes the ith *constellation point* in signal space with every member s_{ij} of the set of N-tuples $\{s_{i0}, s_{i1}, \ldots, s_{iN-1}\}$ representing the projection of $s_i(t)$ along the direction of the jth axis $\varphi_j(t)$.

The length of a vector $\mathbf{s_i}$ in $\mathcal{R}^N$ is

$$||\mathbf{s_i}|| = \sqrt{\sum_{j=0}^{N-1} s_{ij}^2} \tag{2.32}$$

Thus the Euclidean distance between $s_i(t)$ and $s_k(t)$ is basically the distance between two signal constellation points $\mathbf{s_i}$ and $\mathbf{s_k}$

$$d_{ik} = ||\mathbf{s_i} - \mathbf{s_k}|| = \sqrt{\sum_{j=0}^{N-1} (s_{ij} - s_{kj})^2} \tag{2.33}$$

An alternate expression of d_{ik} may be obtained by writing $s_i(t)$ and $s_k(t)$ in their basis representation as given by (2.27) and using the orthonormal properties of the basis functions described by (2.29)

$$d_{ik} = \sqrt{\int_0^{T_s} [s_i(t) - s_k(t)]^2 \, dt} \tag{2.34}$$

The *minimum distance* between a pair of signal constellation points is defined as

$$d_{min} = \min_{i,k}(d_{ik}) \tag{2.35}$$

For any given modulation method, the symbol error probability (SEP) depends on the minimum distance (d_{min}) between the constellation points.

As per the model given in Fig. 2.1, when signal $s_i(t)$ is transmitted, the received signal at demodulator input is given by $r(t) = s_i(t) + n(t)$. Considering the projections n_j

$$n_j = \int_0^{T_s} n(t)\varphi_j(t)dt \tag{2.36}$$

of the noise signal $n(t)$ along the basis functions $\{\varphi_j(t)\}_{j=0}^{N-1}$ and from (2.27) we may

16

rewrite $r(t)$ as

$$r(t) = \sum_{j=0}^{N-1} (s_{ij} + n_j) \varphi_j(t) + n_r(t)$$

$$= \sum_{j=0}^{N-1} r_j \varphi_j(t) + n_r(t)$$

(2.37)

where the coefficients $r_j = s_{ij} + n_j$ form a received signal vector

$$\mathbf{r} = [r_0 \quad r_1 \quad \cdots \quad r_N - 1]^T \in \mathcal{R}^N \tag{2.38}$$

and

$$n_r(t) = n(t) - \sum_{j=0}^{N-1} n_j \varphi_j(t) \tag{2.39}$$

denotes the *remainder* noise. This component of the noise is orthogonal to the signal space and does not contribute in signal demodulation. It may be shown [4] that the N-tuple $\{r_0, r_1, \ldots, r_N - 1\}$ forming $\mathbf{r} \in \mathcal{R}^N$ provides *sufficient statistic* for optimal demodulation. The optimal ML demodulator operates on the minimum distance criteria and decides in favour of the signal vector $\mathbf{s_i}$ that is closest to the received vector $\mathbf{r}$, i.e. it sets $\hat{a}_i = a_i$ whenever

$$||\mathbf{r} - \mathbf{s_i}|| \geq ||\mathbf{r} - \mathbf{s_k}|| \quad ; k \in \{0, 1, \ldots, M - 1\}, k \neq i \tag{2.40}$$

With the vector space representation we can analyze the infinite dimensional functions $s_i(t)$ as vectors $\mathbf{s_i}$ in finite dimensional vector space $\mathcal{R}^N$. This greatly simplifies the analysis of the system performance as well as the derivation of the optimal demodulator design. The demodulator structures with correlators described in Fig. 2.2 may be now redrawn as the one given in Fig. 2.4. The advantage in terms of the reduced number of branches (when $N \leq M$) is clearly visible.

Similarly an alternative to matched filter based demodulator structure in Fig. 2.3 may be thought of in terms of the basis functions. This structure, as shown in Fig. 2.5 makes use of a bank of filters whose impulse responses $\{h_j(t)\}_{j=0}^{N-1}$ are matched

$$h_j(t) = \varphi_j(T_s - t) \quad ; 0 \leq t \leq T_s \tag{2.41}$$

to each of the different basis functions $\{\varphi_j(t)\}_{j=0}^{N-1}$.

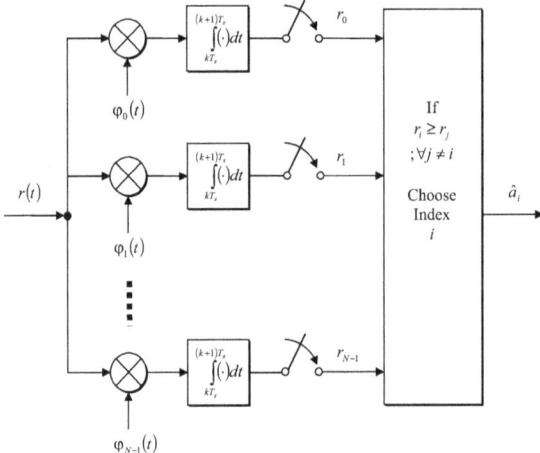

Figure 2.4: Demodulator structure correlated to basis functions.

2.3 Error Probability of Digital Signals

In this section, error performances of different memoryless (linear/ non-linear) modulation methods over AWGN channel have been studied. Both coherent and non-coherent demodulation techniques are considered. The reason for doing this is twofold: first, AWGN channel is a universal channel model for analyzing modulation schemes. When the signal bandwidth is smaller than the channel bandwidth, many wired (telephone channel, wideband coaxial cables) as well as wireless (fixed terrestrial microwave links and fixed satellite links when the weather is good) channels may be modelled as AWGN channel [6]. The second reason is, AWGN is ever present regardless of whether other channel impairments such as multipath fading, interference etc. exist or not. Thus symbol error probability (SEP) or, bit error probability (BEP) of a modulation scheme evaluated in AWGN channel is a lower bound on the error performance. When other channel impairments exist, the system performance will degrade and the extent of degradation may vary for different modulation schemes. The performance in AWGN can serve as a base level in evaluating the degradation and also in evaluating effectiveness of different mitigation techniques.

2.3.1 Binary Modulation

The digital information is assumed to be binary $\{0, 1\}$ and two signals $s_i(t)$; $i = 0, 1$ are needed to represent the binary digits. This type of transmission is called binary signalling and may be viewed as a special case of the general M-ary signalling where $M = 2$. Bi-

18

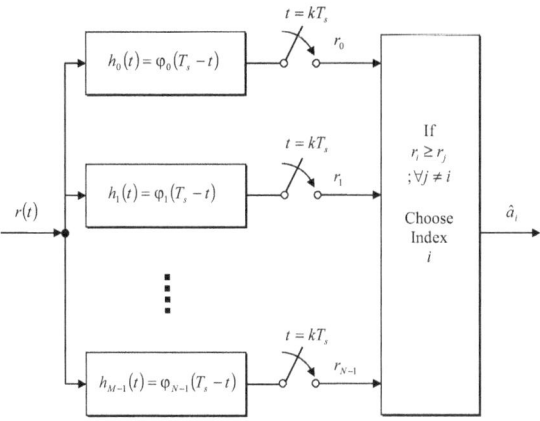

Figure 2.5: Demodulator structure with filters matched to basis functions.

nary modulations were the first to be introduced and still used widely for their simplicity and cost-effectiveness. In the wireless domain they are well suited for the low data rate applications. Binary phase shift keying (BPSK), where two different phases of the carrier are transmitted to differentiate a 0 or 1, is used extensively in wireless local area networks (WLANs) e.g. IEEE 802.11a/g *wireless fidelity* (WiFi) and IEEE 802.16d/e *worldwide interoperability for microwave access* (WiMax). In case of differential phase shift keying (DPSK) a differential coding is used prior to phase modulation which eradicates any possibility of phase reference inversion. This feature makes DPSK more robust against channel impairments and renders it as a modulation choice for IEEE 802.11b and IEEE 802.15.4 wireless personal area network (WPAN) standards. Binary frequency shift keying (BFSK), on the other hand, is more popular in short range or energy constrained wireless services like sensor networks, underwater communications, and is already a part of different standards - IEEE 802.15.1 Bluetooth, IEEE 802.15.3 ultra wide band (UWB) and IEEE 802.15.4 Zigbee. The non-coherent version of BFSK provides significant reduction in receiver complexity and thus sometimes preferred over its coherent counterpart.

Wojner [7] presented a general expression for BEP of binary signalling schemes in terms of complementary incomplete gamma function $\Gamma(\cdot, \cdot)$ [8, (8.350.2)] namely,

$$P_{e,b} = \frac{\Gamma(\beta, \alpha\eta_b)}{2\Gamma(\beta)} \tag{2.42}$$

where,

$$\beta = \begin{cases} 1/2 & \text{for coherent detection} \\ 1 & \text{for differential/non-coherent detection} \end{cases}$$

and

$$\alpha = \begin{cases} 1/2 & \text{for orthogonal FSK} \\ 1 & \text{for antipodal PSK} \end{cases}$$

whereas $\eta_b = E_b/N_0$ is the signal to noise ratio (SNR) per bit.

Using the set of relations $\Gamma(1/2, z^2) = \sqrt{\pi}\operatorname{erfc}(z)$ [9, (6.5.17)] and $\Gamma(1, z) = \exp(-z)$ [8, (8.352.2)], the error probabilities can be written in the commonly available form. For example, the BEP for coherent BPSK is

$$P_{e,b} = \frac{1}{2}\operatorname{erfc}(\sqrt{\eta_b}) \tag{2.43}$$

where $\operatorname{erfc}(\cdot)$ denotes complementary error function [9, (7.1.2)]. Further, using Craig's representation for $\operatorname{erfc}(\cdot)$ [10, (10)], the BEP may also be written in the form of a finite range integral as

$$P_{e,b} = \frac{1}{\pi} \int_0^{\frac{\pi}{2}} \exp\left(-\frac{\eta_b}{\sin^2 \theta}\right) d\theta \tag{2.44}$$

This form is particularly suitable for evaluating average BEP in wireless fading channels where the BEP expression in (2.44) needs to be integrated over the fading PDF. More on this is detailed in Chapter 3 during a discussion about SEP calculation through moment generating function (MGF) approach.

For coherent BFSK the BEP in closed form is

$$P_{e,b} = \frac{1}{2}\operatorname{erfc}\left(\sqrt{\frac{\eta_b}{2}}\right) \tag{2.45}$$

and the corresponding finite range integral representation may be written as

$$P_{e,b} = \frac{1}{\pi} \int_0^{\frac{\pi}{2}} \exp\left(-\frac{\eta_b}{2\sin^2 \theta}\right) d\theta \tag{2.46}$$

For DPSK and non-coherent BFSK (ncBFSK), however, the error rates are available in simpler form, involving only the elementary exponential function. For DPSK the BEP is

$$P_{e,b} = \frac{1}{2}\exp(-\eta_b) \tag{2.47}$$

while for ncBFSK the BEP becomes

$$P_{e,b} = \frac{1}{2} \exp\left(-\frac{\eta_b}{2}\right) \tag{2.48}$$

There is yet another binary modulation of interest, differentially coded and coherently demodulated BPSK, popularly referred to DBPSK. DBPSK is used in 802.11b compliant WLANs for transmitting data at 1 Mbps. The corresponding error rate is given by [6, (4.12)]

$$P_{e,b} = \text{erfc}(\sqrt{\eta_b}) - \frac{1}{2}\text{erfc}^2(\sqrt{\eta_b}) \tag{2.49}$$

Using the alternate representation of $Q^2(\cdot)$ [11] and the relation $Q(z) = (1/2)\text{erfc}(z/\sqrt{2})$, where $Q(\cdot)$ is the Gaussian probability integral, an integral form for BEP of DBPSK can also be derived as

$$
\begin{aligned}
P_{e,b} &= \frac{2}{\pi}\left[\int_0^{\frac{\pi}{2}} \exp\left(-\frac{\eta_b}{\sin^2\theta}\right) d\theta - \int_0^{\frac{\pi}{4}} \exp\left(-\frac{\eta_b}{\sin^2\theta}\right) d\theta\right] \\
&= \frac{2}{\pi}\int_{\frac{\pi}{4}}^{\frac{\pi}{2}} \exp\left(-\frac{\eta_b}{\sin^2\theta}\right) d\theta
\end{aligned}
\tag{2.50}
$$

Writing the same in terms of $\pi/2 - \theta$ instead of θ yields the following form

$$P_{e,b} = \frac{2}{\pi}\int_0^{\frac{\pi}{4}} \exp\left(-\frac{\eta_b}{\cos^2\theta}\right) d\theta \tag{2.51}$$

When closed-form expressions are to be derived, to avoid unnecessary computational complexity, the second term in (2.49) is sometimes omitted. The remaining part

$$P_{e,b} \leq \text{erfc}(\sqrt{\eta_b}) = \frac{2}{\pi}\int_0^{\frac{\pi}{2}} \exp\left(-\frac{\eta_b}{\sin^2\theta}\right) d\theta \tag{2.52}$$

serves as a upper bound which becomes increasingly tight at large η_b values.

2.3.2 MPSK

The data-centric networks are trying hard to cope up with new multimedia services introduced every other day. For smooth functioning of these applications very high capacity transmission is required, with the latest application demanding still higher data rate than its predecessor. Current wireless transceiver designs are resorting to M-ary modulation formats to achieve simultaneous transmission of multiple information bits. Traditionally M-ary phase shift keying (MPSK) has been preferred [12] as it can realize spectral efficiency at the expense of power efficiency and work well when the system is bandwidth constrained. The most commonly used digital modulation in satellite communication is MPSK [13, 14, 15]. MPSK is also used in deep space [16] and optical wireless [17] applications.

With reference to the general expression of digitally modulated signal given in (2.15), for MPSK systems the signal amplitude is constant ($A_i = \sqrt{2E_s/T_s}$) while $f_i = 0$. The phase of the carrier ϕ_i is allowed to take on one of the M possible values

$$\phi_i = \frac{2\pi i}{M} \quad ; i \in \{0, 1, \dots, M-1\} \tag{2.53}$$

to represent M distinct symbols of a binary sequence of length $n = \log_2 M$. Thus M-possible signals that would be transmitted during each signaling interval of duration T_s is

$$s_i(t) = \sqrt{\frac{2E_s}{T_s}} \cos\left(2\pi f_c t + \frac{2\pi i}{M}\right) \quad ; i \in \{0, 1, \dots, M-1\} \tag{2.54}$$

Utilizing the set of values $\{A_i, f_i, \phi_i\}$ for MPSK the equivalent baseband representation may be obtained from (2.24) as

$$\tilde{s}_i(t) = \sqrt{\frac{2E_s}{T_s}} \exp\left(j2\pi \frac{i}{M}\right) \quad ; i \in \{0, 1, \dots, M-1\} \tag{2.55}$$

Recognizing that each $s_i(t)$ may be written in terms of two orthogonal basis functions

$$\varphi_0(t) = \sqrt{\frac{2}{T_s}} \cos(2\pi f_c t) \tag{2.56a}$$

$$\varphi_1(t) = \sqrt{\frac{2}{T_s}} \sin(2\pi f_c t) \tag{2.56b}$$

the coordinates of the signal constellation points can be calculated as

$$s_{i0} = \int_0^{T_s} s_i(t)\varphi_0(t)dt = \sqrt{E_s} \cos\left(\frac{2\pi i}{M}\right) \tag{2.57a}$$

$$s_{i1} = \int_0^{T_s} s_i(t)\varphi_1(t)dt = \sqrt{E_s} \sin\left(\frac{2\pi i}{M}\right) \tag{2.57b}$$

which, when plotted, gives a two dimensional circular constellation diagram. Each constellation point is equally spaced on a circle of radius $\sqrt{E_s}$ centered at the origin and maintaining a angular distance of $2\pi/M$ from its adjacent constellation point. Fig. 2.6 shows constellation diagram of a 8-ary PSK with Gray encoding. *Gray coding* assigns n-bit groups with only one bit difference to two adjacent symbols in the constellation. When a symbol is wrongly decoded it is more likely that the signal is detected as the adjacent signal on the constellation, thus only one out of the $n = \log_2 M$ input bits is in

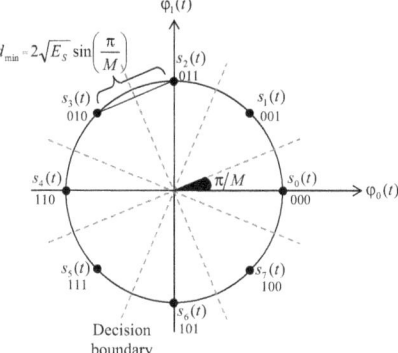

Figure 2.6: Constellation diagram for MPSK ($M = 8$) with Gray coding.

error. Therefore, the average BEP for MPSK can be approximately related to SEP as

$$P_{e,b} \approx \frac{P_e}{\log_2 M} = \frac{P_e}{n} \tag{2.58}$$

Since the MPSK signal set has only two basis functions, the simplest demodulator is the one that uses two correlators ($N = 2$) as depicted in Fig. 2.7. The receiver decides whether the received phase θ is within $\pm \pi/M$ of θ_i.

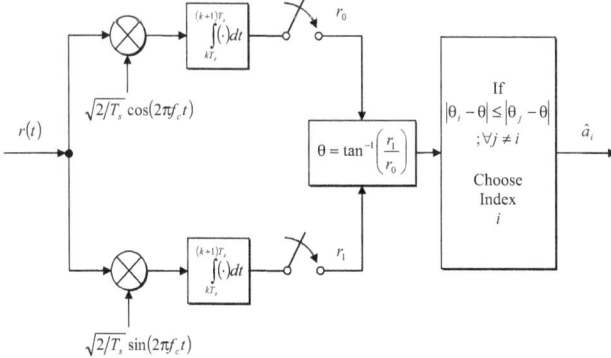

Figure 2.7: Coherent MPSK demodulator using two correlators.

For calculation of error probability, without losing generality, we may assume that

$s_0(t)$ corresponding to zero phase shift, is being transmitted. Under this assumption the received signal is

$$r(t) = s_0(t) + n(t) \tag{2.59}$$

However the actual noise affecting the decision process is a narrowband signal centered around f_c. Using the quadrature representation of noise signal (2.25) the outputs of the in-phase and quadrature correlator branches may be expressed as

$$
\begin{aligned}
r_0 &= \int_0^{T_s} \left[\sqrt{\frac{2E_s}{T_s}} + \tilde{n}_I(t) \right] \cos(2\pi f_c t) \sqrt{\frac{2}{T_s}} \cos(2\pi f_c t) dt \\
&= \sqrt{E_s} + n_c
\end{aligned} \tag{2.60a}
$$

$$
\begin{aligned}
r_1 &= \int_0^{T_s} \tilde{n}_Q(t) \sin(2\pi f_c t) \sqrt{\frac{2}{T_s}} \sin(2\pi f_c t) dt \\
&= n_s
\end{aligned} \tag{2.60b}
$$

where the RVs $n_c, n_s \sim \mathcal{N}(0, N_0/2)$. Naturally, $r_0 \sim \mathcal{N}(\sqrt{E_s}, N_0/2)$ while $r_1 \sim \mathcal{N}(0, N_0/2)$ and the joint PDF of r_0, r_1 can be written as

$$f_{r_0, r_1}(r_0, r_1) = \frac{1}{\pi N_0} \exp\left[-\frac{(r_0 - \sqrt{E_s})^2 + r_1^2}{N_0} \right] \tag{2.61}$$

Let us define the amplitude and phase of the variable $R = r_0 + jr_1$ as r and θ where $r = \sqrt{r_0^2 + r_1^2}$ and $\theta = \tan^{-1}(r_1/r_0)$. Through the following transformation of RVs $r_0 = r\cos\theta$ and $r_1 = r\sin\theta$, the joint PDF of r and θ can be obtained as [18, 19]

$$
\begin{aligned}
f_{r,\theta}(r, \theta) &= f_{r_0, r_1}(r\cos\theta, r\sin\theta)|J| \\
&= \frac{r}{\pi N_0} \exp\left[-\frac{(r\cos\theta - \sqrt{E_s})^2 + (r\sin\theta)^2}{N_0} \right]
\end{aligned} \tag{2.62}
$$

where $|J|$ is the Jacobian of the transformation

$$|J| = \left| \frac{\partial(r_0, r_1)}{\partial(r, \theta)} \right| = \left| \begin{matrix} \partial r_0/\partial r & \partial r_0/\partial\theta \\ \partial r_1/\partial r & \partial r_1/\partial\theta \end{matrix} \right| = \left| \begin{matrix} \cos\theta & -r\sin\theta \\ \sin\theta & -r\cos\theta \end{matrix} \right| = r \tag{2.63}$$

The marginal densities of both r and θ can be calculated by integrating the joint PDF $f_{r,\theta}(r, \theta)$ with respect to the other variable. For the amplitude part, it can be easily shown

[18] that $f_r(r)$ follows a Rician Distribution. For the phase

$$
\begin{aligned}
f_\theta(\theta) &= \int_{r=0}^{\infty} f_{r,\theta}(r,\theta)dr \\
&= \frac{1}{\pi N_0} \exp\left(-\frac{E_s \sin^2 \theta}{N_0}\right) \\
&\quad \times \int_0^{\infty} r \exp\left[-\frac{(r - \sqrt{E_s} \cos\theta)^2}{N_0}\right] dr
\end{aligned}
\tag{2.64}
$$

Substituting $t = (r - \sqrt{E_s} \cos\theta)/\sqrt{N_0}$ we can write (2.64) as

$$
f_\theta(\theta) = \frac{\exp(-\eta \sin^2 \theta)}{\pi} \left[\int_{-\sqrt{\eta}\cos\theta}^{\infty} t \exp(-t^2)dt \right.
$$
$$
\left. + \sqrt{\eta} \cos\theta \int_{-\sqrt{\eta}\cos\theta}^{\infty} \exp(-t^2)dt \right]
\tag{2.65}
$$

where $\eta = E_s/N_0$ is the SNR per symbol. After evaluation of each term of the integral in (2.65) we have

$$
f_\theta(\theta) = \frac{\exp(-\eta)}{2\pi} \left[1 + \sqrt{4\pi\eta} \cos\theta \exp(\eta \cos^2 \theta) Q(-\sqrt{2\eta} \cos\theta) \right]
\tag{2.66}
$$

as obtained by Stein and Jones [20].

As $s_0(t)$ is the signal transmitted, a correct decision is made if $-\pi/M \le \theta \le \pi/M$. Therefore the probability of symbol error is given by

$$
P_e = 1 - \int_{-\pi/M}^{\pi/M} f_\theta(\theta)d\theta
\tag{2.67}
$$

With the help of the approximation $Q(z) \approx 1/(z\sqrt{2\pi}) \exp(-z^2/2)$ for $z \gg 1$ and using the relation $Q(-z) = 1 - Q(z)$, evaluation of the above integration becomes simpler when the integrand $f_\theta(\theta)$ given in (2.66) is expressed in a more convenient form

$$
\begin{aligned}
f_\theta(\theta) &\approx \frac{\exp(-\eta)}{2\pi} [1 + \sqrt{4\pi\eta} \cos\theta \exp(\eta \cos^2 \theta) \\
&\qquad \times \{ 1 - 1/(\exp(\eta \cos^2 \theta)\sqrt{4\pi\eta} \cos\theta) \}] \\
&\approx \sqrt{\frac{\eta}{\pi}} \exp(-\eta \sin^2 \theta) \cos\theta
\end{aligned}
\tag{2.68}
$$

Substituting $u = \sqrt{\eta} \sin\theta$ and performing the integration, the probability of error becomes

$$
P_e \approx 1 - \frac{1}{\sqrt{\pi}} \int_{-\sqrt{\eta}\sin(\pi/M)}^{\sqrt{\eta}\sin(\pi/M)} \exp(-t^2)dt = \mathrm{erfc}\left[\sqrt{\eta} \sin\left(\frac{\pi}{M}\right)\right]
\tag{2.69}
$$

Equation (2.69) serves as an union upper bound for the SEP of coherent MPSK. Using alternate finite range integral expression of $\mathrm{erfc}(\cdot)$ the bound becomes

$$P_e \leq \frac{2}{\pi} \int_0^{\pi/2} \exp\left[-\frac{\eta \sin^2(\pi/M)}{\sin^2\theta}\right] d\theta \tag{2.70}$$

which is an alternate form of upper bound as described by Sun and Reed [21, (10)]

$$P_e \leq \frac{1}{\pi} \int_{-\pi/2}^{\pi/2} \exp\left[-\frac{\eta \sin^2(\pi/M)}{\cos^2\theta}\right] d\theta \tag{2.71}$$

By noting the symmetry of the integral in (2.71) about $\theta = 0$, it is easy to verify that for both the ranges $0 \leq \theta \leq \pi/2$ and $-\pi/2 \leq \theta \leq 0$ (2.71) reduces to (2.70) when the former is written in terms of $\pi/2 - \theta$ instead of θ. The exact formula for the symbol error probability of MPSK, however, was first given by Pawula [22, (2)]

$$P_e = \frac{1}{\pi} \int_0^{\pi-\pi/M} \exp\left[-\frac{\eta \sin^2(\pi/M)}{\sin^2\theta}\right] d\theta \tag{2.72}$$

For large η values the difference between (2.70) and (2.72) becomes negligibly small.

Quadrature phase shift keying (QPSK) is a special member of the general MPSK family where $M = 4$. The modulation was used in second generation (2G) code division multiple access (CDMA) cellular standard IS-95 and currently being used in forward link of third generation (3G) standards like wideband CDMA (WCDMA) and cdma2000. QPSK is standardized as one of the modulation schemes for IEEE 802.11a/g WLAN and also widely used in mobile television standards (DVB-H, MediaFLO, ISDB-T) [23]. A QPSK signal comprises of two BPSK signals in quadrature and when either of these signals are wrongly demodulated the corresponding QPSK symbol becomes erroneous. Thus the SEP for QPSK may be found from (2.43) as

$$
\begin{aligned}
P_e &= 1 - \left[1 - P_{e,b(BPSK)}\right]^2 \\
&= \mathrm{erfc}(\sqrt{\eta_b}) - \frac{1}{4}\mathrm{erfc}^2(\sqrt{\eta_b})
\end{aligned}
\tag{2.73}
$$

Noting that the symbol energy E_s for QPSK is two times of the bit energy E_b, the SEP expression may also be written in terms of the SNR per symbol η as

$$P_e = \mathrm{erfc}\left(\sqrt{\frac{\eta}{2}}\right) - \frac{1}{4}\mathrm{erfc}^2\left(\sqrt{\frac{\eta}{2}}\right) \tag{2.74}$$

Eq. (2.74) describes not only the SEP of QPSK but it also gives SEP values of some QPSK variants like offset QPSK (OQPSK) and minimum shift keying (MSK). In OQPSK, by offsetting the timing of the odd and even bits by T_b or $T_s/2$, it is made sure that the in-phase and quadrature components will never change at the same time. MSK is much similar to OQPSK except the fact that it uses sinusoidal pulse shaping instead of rectangular pulse. For larger η values the contribution of the second term almost vanishes and thus an upper bound may be obtained by omitting the second term in (2.74) as

$$P_e \leq \text{erfc}\left(\sqrt{\frac{\eta}{2}}\right) \tag{2.75}$$

which is nothing but the bound given in (2.69) for $M = 4$. The corresponding BEP can be calculated from (2.75) and $\eta = 2\eta_b$ using (2.58)

$$P_{e,b} \approx \frac{1}{2}\text{erfc}(\sqrt{\eta_b}) \tag{2.76}$$

From (2.76) and (2.43) it is clear that the BEP values for BPSK ($M = 2$) and QPSK ($M = 4$) are same but a QPSK symbol can carry two bits compared to BPSK where only single bit is carried every time. Thus switching to QPSK from BPSK doubles the bandwidth efficiency without any penalty in the power domain. However, in the MPSK family, this typical behaviour is limited up to $M = 4$ only. When the transmitter power is fixed, i.e. at a given η_b value, both bandwidth efficiency and BEP increases with M for $M > 4$.

2.3.3 MDPSK

Implementation and operation of a coherent receiver is quite complex due to the task of realizing a carrier demodulation reference that is perfectly synchronized in phase and frequency with the received signal. A carrier recovery loop is generally used for the purpose which involves acquisition, tracking, lock detection, false lock prevention etc. [24] and thereby increases the complexity of the demodulator manifold. Aside from implementation considerations, the transmission environment is heavily degraded in wireless multipath fading channels and parameter estimation for carrier synchronization is in general a difficult task. There is no doubt that coherent demodulation yields better error performance than differential detection. However, differential detection provides an attractive alternative when simplicity and robustness are prime issues.

For an M-ary differential phase shift keying (MDPSK) system, the signal transmitted in the kth transmission interval is

$$s_i(t) = \sqrt{\frac{2E_s}{T_s}}\cos(2\pi f_c t + \phi_k) \quad ; kT_s \leq t \leq (k+1)T_s \tag{2.77}$$

where the current phase ϕ_k is related to the previously transmitted phase ϕ_{k-1} as

$$\phi_k = \Delta\phi + \phi_{k-1} \tag{2.78}$$

27

The differential phase is allowed to take on one of the M possible values

$$\Delta\phi = \theta_i = \frac{2\pi i}{M} \quad ; i \in \{0, 1, \ldots, M-1\} \tag{2.79}$$

The equivalent baseband representation is thus

$$\tilde{s}_i(t) = \sqrt{\frac{2E_s}{T_s}} \exp(j\phi_k) \quad ; i \in \{0, 1, \ldots, M-1\} \tag{2.80}$$

where ϕ_k is given by (2.78) and (2.79).

Like MPSK, demodulator for MDPSK may also be realized with only two ($N = 2$) basis functions

$$\varphi_0(t) = \sqrt{\frac{2}{T_s}} \cos(2\pi f_c t) \tag{2.81a}$$

$$\varphi_1(t) = \sqrt{\frac{2}{T_s}} \sin(2\pi f_c t) \tag{2.81b}$$

and the corresponding structure is shown in Fig. 2.8. It is obvious that, just like MPSK,

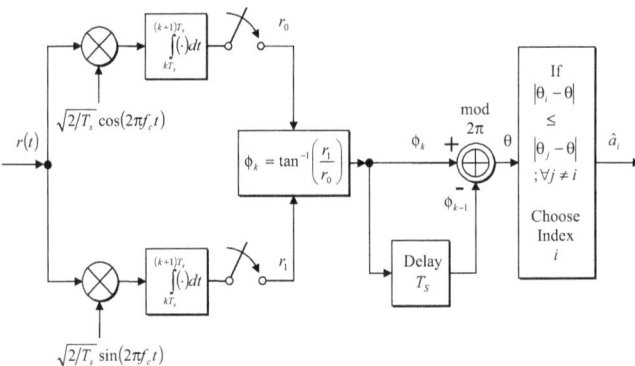

Figure 2.8: MDPSK demodulator using two correlators.

the decision regions are defined as wedges uniformly spaced around a circle. If signal $s_i(t)$ is transmitted, a correct decision is made when the decision variable $\theta = (\phi_k -$

$\phi_{k-1})\mathrm{mod}(2\pi)$ falls in the region $\theta_i - \pi/M \leq \theta \leq \theta_i + \pi/M$

$$
\begin{aligned}
P(C|s_i(t)) &= \Pr\left[|\theta - \theta_i| \leq \frac{\pi}{M}\right] \\
&= \int_{-\pi/M}^{\pi/M} f_\chi(\chi)d\chi
\end{aligned}
\tag{2.82}
$$

where $\chi = \theta - \theta_i$. The PDF $f_\chi(\chi)$ was shown to be[24]

$$
\begin{aligned}
f_\chi(\chi) &= \frac{1}{2\pi}\int_0^{\pi/2} \sin\alpha[1 + \eta(1 + \cos\chi\sin\alpha)] \\
&\quad \times \exp[-\eta(1 - \cos\chi\sin\alpha)]d\alpha
\end{aligned}
\quad ; |\chi| \leq \pi
\tag{2.83}
$$

Since this PDF is independent of k, the average SEP is

$$
\begin{aligned}
P_e &= 1 - \int_{-\pi/M}^{\pi/M} f_\chi(\chi)d\chi \\
&= \int_{-\pi}^{\pi} f_\chi(\chi)d\chi - \int_{-\pi/M}^{\pi/M} f_\chi(\chi)d\chi \\
&= 2\int_{\pi/M}^{\pi} f_\chi(\chi)d\chi
\end{aligned}
\tag{2.84}
$$

where the simplification in the last step was possible due to symmetry of $f_\chi(\chi)$ about $\chi = 0$. Without losing generality we may assume that $\theta_i = 0$ and this enables us to write (2.84) as

$$
P_e = 2\int_{\pi/M}^{\pi} f_\theta(\theta)d\theta
\tag{2.85}
$$

Solution to the integral given in (2.85) was carried out by Pawula et al. [25] rendering the SEP in the following form

$$
P_e = \frac{\sin(\pi/M)}{2\pi}\int_{-\pi/2}^{\pi/2} \frac{\exp\left[-\eta\{1 - \cos(\pi/M)\cos\theta\}\right]}{1 - \cos(\pi/M)\cos\theta}d\theta
\tag{2.86}
$$

involving a single integral of elementary functions. Many research papers, including the one by Sun and Reed [21, (3)], used this form of SEP for MDPSK. Another compact and accurate formula for SEP was reported later on

$$
P_e = \frac{1}{\pi}\int_0^{\pi - \pi/M} \exp\left[-\frac{\eta\sin^2(\pi/M)}{1 + \cos(\pi/M)\cos\theta}\right]d\theta
\tag{2.87}
$$

by Pawula [22, (3)] in a separate paper. It is interesting to note the similarity between (2.72) and (2.87). Also an approximated SEP formula is sometimes used for reducing

29

complexity in analysis

$$P_e \approx \text{erfc}\left[\sqrt{\eta}\sin\left(\frac{\pi}{\sqrt{2}M}\right)\right] \tag{2.88}$$

which was developed by Arthurs and Dym [26] and having a form similar to that in (2.69).

In the MDPSK family, a special case of interest is $\pi/4$-shifted differential quadrature phase shift keying ($\pi/4$-DQPSK), i.e. MDPSK with $M = 4$, due to its wide applicability in cellular communication systems. From the inception of cellular radio the modulation format $\pi/4$-DQPSK was closely associated with such systems for its robustness towards multipath fading and interference. The pioneer U.S. and Japanese CDMA 2G mobile radio standards like USDC/IS-136, PDC etc. operating in 900 MHz band incorporated $\pi/4$-DQPSK. For concurrent *cordless telephone* (CT) and *wireless in local loop* (WLL) type applications in 1.8/1.9 GHz band the same modulation was included in U.S. PACS or Japanese PHS system [27]. In WLAN domain, the popular DSSS mode in IEEE 802.11 had the provision of realizing 2 Mbps data rate using $\pi/4$-DQPSK [28]. DQPSK capability is also made mandatory in the physical layer specifications for 2.4 GHz, 22 Mbps IEEE 802.15.3 WPAN [29]. Some other applications of DQPSK include terrestrial trunked radio (TETRA) for emergency, military and transportation services, land mobile satellite [30], digital audio (EU 147) and video (T-DMB) broadcasting, space time block codes [31], turbo coding [32], and mobile ad hoc networks [33].

Quadrature modulations give the best compromise between bandwidth efficiency and BEP performance, out of which $\pi/4$-DQPSK is preferred to its conventional counterparts QPSK and OQPSK. The reason being the fact that unlike QPSK, $\pi/4$-DQPSK does not suffer from 180^0 envelope variation (maximum phase shift is restricted to $\pm135^0$) causing amplitude dips at bit transitions; which become even more pronounced in presence of pulse shaping, i.e. when non-rectangular basis pulses are used. In addition, unlike OQPSK, it can be differentially detected, which removes complex costly phase coherent operations in the receiver and makes it suitable for fading channel applications.

For $\pi/4$-DQPSK the phase change between successive symbols are maintained according to the rule

$$\Delta\phi = \theta_i = \left(\frac{\pi}{4} + \frac{\pi i}{2}\right)\text{mod}(2\pi) \tag{2.89}$$

where $i \in \{0,1,2,3\}$ and $\Delta\phi \in \{\pm\pi/4, \pm3\pi/4\}$. The index i gives symbol number which depends on gray coded input bit pattern as shown in Table 2.1.

As seen from Fig. 2.9, $\pi/4$-DQPSK constellation can be described as a superposition of two QPSK signal constellations offset by $\pi/4$ relative to each other (denoted with circles and squares). The two signal constellations are used alternately from symbol to symbol. Every symbol consists of two binary digits. For odd numbered symbols we use

30

Input Bits	Gray Coded Input Bits	Symbol (i)	Phase Change in Radian ($\Delta\phi = \theta_i$)		
00	00	0	$\pi/4 + 0$	$=$	$+\pi/4$
01	01	1	$\pi/4 + \pi/2$	$=$	$+3\pi/4$
10	11	2	$\pi/4 + \pi$	$=$	$-3\pi/4$
11	10	3	$\pi/4 + 3\pi/2$	$=$	$-\pi/4$

Table 2.1: Description of $\pi/4$-DQPSK symbol encoding.

QPSK with phase shift 0 and for even numbered symbols we use QPSK with phase shift $\pi/4$. Thus the signal may have one of the eight possible phases $n\pi/4, n = 0, 1, \ldots, 7$ but successive symbols differ in phase by only $\pi/4$ or, $3\pi/4$ [34]. All possible phase changes are shown with arrows in Fig. 2.9.

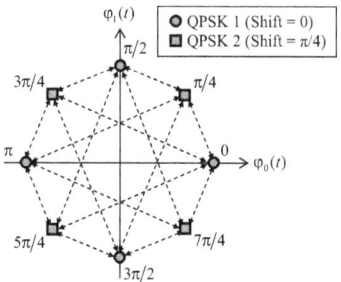

Figure 2.9: $\pi/4$-DQPSK constellation diagram.

A demodulator with differential detector for $\pi/4$-DQPSK has been depicted in Fig. 2.10. Correlators are replaced by the matched filters matched to the basis functions, i.e. $h_i(t) = \varphi_i(T_s - t); i \in \{0, 1\}$. This structure avoids costly and complex carrier synchronization circuitry and helps in realizing simpler and less power consuming demodulator which is ideal for battery driven wireless transceivers. As evident from Fig. 2.10, at any sampling instant $t = kT_s$ the outputs of matched filters may be written as

$$r_{0,k} = \sqrt{E_s} \cos\theta_k \tag{2.90a}$$
$$r_{1,k} = \sqrt{E_s} \sin\theta_k \tag{2.90b}$$

where phase of the current signal θ_k is related to the phase of the previous signal as $\theta_k = \theta + \theta_{k-1}$. Substituting this relation in (2.90) and making use of the expansion

31

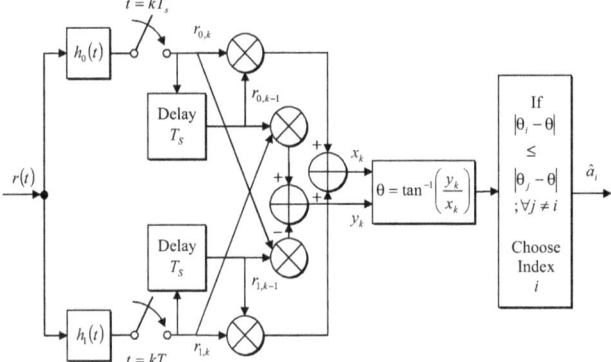

Figure 2.10: $\pi/4$-DQPSK demodulator with differential detector.

formulas of $\cos(A + B)$ and $\sin(A + B)$ we can write (2.90) as

$$r_{0,k} = r_{0,k-1} \cos \theta - r_{1,k-1} \sin \theta \tag{2.91a}$$

$$r_{1,k} = r_{0,k-1} \sin \theta + r_{1,k-1} \cos \theta \tag{2.91b}$$

Referring to Fig. 2.10 once again and noting that $r_{0,k}^2 + r_{1,k}^2 = E_s; \forall k$ we find that the output of the adders are

$$x_k = r_{0,k} r_{0,k-1} + r_{1,k} r_{1,k-1} = E_s \cos \theta \tag{2.92a}$$

$$y_k = r_{1,k} r_{0,k-1} - r_{0,k} r_{1,k-1} = E_s \sin \theta \tag{2.92b}$$

The incremental phase difference θ is then found by $\theta = \tan^{-1}(y_k/x_k)$ and a decision device, following a ML rule, converts the information to raw output bits according to Table 2.1.

Assuming complete CSI available at the receiver, the average BEP for $\pi/4$-DQPSK modulation with differential detection in an AWGN channel was derived by Miller and Lee [35]

$$
\begin{aligned}
P_{e,b} \quad = \quad & \frac{1}{2} \left[1 - Q \left(\sqrt{\eta_b(2 + \sqrt{2})}, \sqrt{\eta_b(2 - \sqrt{2})} \right) \right. \\
& \left. + Q \left(\sqrt{\eta_b(2 - \sqrt{2})}, \sqrt{\eta_b(2 + \sqrt{2})} \right) \right]
\end{aligned}
\tag{2.93}
$$

where $Q(\cdot, \cdot)$ denotes Marcum's Q function of first order, i.e. the special case of generalized Marcum's Q function $Q_m(\cdot, \cdot)$ for $m = 1$, also written as $Q_1(\cdot, \cdot)$. Using infinite series expressions of Marcum's Q function [36], the BEP can also be written as,

$$P_{e,b} = Q\left(\sqrt{\eta_b(2 - \sqrt{2})}, \sqrt{\eta_b(2 + \sqrt{2})}\right) - \frac{1}{2}\exp(-2\eta_b)I_0\left(\sqrt{2}\eta_b\right) \tag{2.94}$$

which is same as [2, (5.2.70)], the error probability expression for four-phase Gray-coded DPSK modulation. $I_m(\cdot)$ denotes the mth-order modified Bessel function of the first kind. Under the assumption that the parameters a and b of $Q_1(a,b)$ are proportional to the square root of SNR, i.e. in the form $Q_1(\alpha\sqrt{\eta}, \beta\sqrt{\eta})$, we have the following alternate representation [37]

$$Q_1(a,b) = \begin{cases} \dfrac{1}{2\pi}\displaystyle\int_{-\pi}^{\pi} \dfrac{f_1(a,b,\theta)}{f_2(a,b,\theta)}\exp\left(-\dfrac{f_2(a,b,\theta)}{2}\right)d\theta \\ \qquad\qquad\qquad\qquad\qquad \text{for } b > a \geq 0 \\ 1 + \dfrac{1}{2\pi}\displaystyle\int_{-\pi}^{\pi} \dfrac{f_1(a,b,\theta)}{f_2(a,b,\theta)}\exp\left(-\dfrac{f_2(a,b,\theta)}{2}\right)d\theta \\ \qquad\qquad\qquad\qquad\qquad \text{for } a > b \geq 0 \end{cases} \tag{2.95}$$

where $f_1(a,b,\theta) = b^2 + ab\sin\theta$ and $f_2(a,b,\theta) = b^2 + 2ab\sin\theta + a^2$. Using (2.95) the error probability in (2.93) can be expressed as

$$P_{e,b} = \frac{1}{4\pi}\int_{-\pi}^{\pi} \frac{\exp[-\eta_b(2 + \sqrt{2}\sin\theta)]}{\sqrt{2} + \sin\theta}d\theta \tag{2.96}$$

The integral in (2.96) contains $\sin\theta$ whose value within the range $0 \leq \theta \leq \pi$ is positive and symmetric about $\theta = \pi/2$, i.e. $\sin(\pi - \theta) = \sin\theta$. Noting a similar symmetry in the range $-\pi \leq \theta \leq 0$ about $\theta = -\pi/2$, the integral range may be shrinked to $-\pi/2$ to $\pi/2$ while doubling its integrand. Further writing the same in terms of $\pi/2 + \theta$ instead of θ we obtain the following form

$$P_{e,b} = \frac{1}{2\pi}\int_0^{\pi} \frac{\exp[-\eta_b(2 - \sqrt{2}\cos\theta)]}{\sqrt{2} - \cos\theta}d\theta \tag{2.97}$$

as devised by Tellambura and Bhargava [38, (3)] and subsequently reported by Simon and Alouini [39, (28)]. The single integral representation is quite useful while calculating the BEP through MGF method. Another formula for $\pi/4$-DQPSK BEP was reported later on

$$P_{e,b} = \frac{1}{4\pi}\int_{-\pi}^{\pi} \exp\left(-\frac{\sqrt{2}\eta_b}{\sqrt{2} + \sin\theta}\right)d\theta \tag{2.98}$$

involving single integral but having much simpler form compared to (2.97). The BEP expression may also be approximated from (2.86) by putting $M = 4$, $\eta_b = \eta/2$ and noting that $P_{e,b} \approx P_e/2$ as demonstrated in (2.58) assuming Gray encoding. However, the approximation holds good for large SNR values only.

2.3.4   Coherent MFSK

There is a wide range of wireless communication systems - deep space, sensor networks, ultra-wideband (UWB), or, underwater communications - that operate in the low-power regime where power consumption rather than bandwidth is the limiting factor. In such cases orthogonal signaling schemes like M-ary frequency shift keying (MFSK) is a better choice. In fact the capacity results show [40] that it is possible to make the error probabilities of orthogonal signaling arbitrarily small when the constellation size M goes to infinity as long as the normalized SNR is greater than the Shannon's limit, i.e. -1.6 dB. For a given SNR in an AWGN channel, use of coherent MFSK results in an SEP which is lower than the SEP of all other possible M-ary modulation schemes for $M > 2$. But one has to remember that this power advantage comes at the expense of asymptotical increment in bandwidth.

MFSK modulation is widely applied in transmitter energy limited space communication systems such as deep space probes [41], satellites, space telemetry etc. and is suitable for hand-held satellite terminals which require low-complexity low-cost receiver structures [42]. Cellular wireless terminals are generally battery-driven devices and should be handy which restrict their capability of increasing transmission power; especially in the reverse link. In addition, small scale multipath fading in wireless channels contributes to further power penalty. This inspired some researchers [43, 44] to investigate whether M-ary orthogonal signaling is a suitable option for CDMA or not. MFSK is also useful for low-power, short-range applications; may it be a sensor network or, an UWB piconet [45]. Energy efficient operation is vital in sensor networks to maintain a reasonable lifetime. On the other hand, power efficiency of high dimensional orthogonal modulation makes it very attractive for UWB system design. It is also quite natural to associate MFSK with frequency division multiplexing (FDM), frequency diversity and frequency-hop spread spectrum (FHSS) systems [46, 47]. With the advent of multiple tone (MT) modulation scheme MT-MFSK [48] for bandwidth constraint systems and multicarrier (MC) technique MC-MFSK [49] for future high-speed wireless systems, MFSK is gaining more attention of modern researchers and engineers.

In any M-ary modulation the input binary stream is divided into n-tuples of $n = \log_2 M$ bits where M is the constellation size. If the modulator is following an MFSK scheme then it converts every such n bit message to one of the M possible signals differing in frequency. Expression of the ith signal may be derived by setting $A_i = \sqrt{2E_s/T_s}$, $\phi_i = $

$\phi; \forall i$ and $f_i = hi/(2T_s)$ in (2.15) as

$$s_i(t) = \sqrt{\frac{2E_s}{T_s}} \cos\left[2\pi\left(f_c + \frac{ih}{2T_s}\right)t\right] \tag{2.99}$$

where, without losing generality, the initial phase ϕ is assumed to be 0. In order to maintain orthogonality h should be an integer, i.e. the minimum separation between two adjacent frequencies should be $1/(2T_s)$ [2, 6]. From (2.24) it is easy to verify that the complex envelope of the modulated signal is

$$\tilde{s}_i(t) = \sqrt{\frac{2E_s}{T_s}} \exp\left(jh\pi\frac{it}{T_s}\right) \tag{2.100}$$

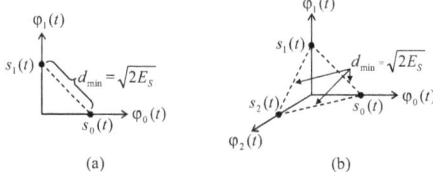

Figure 2.11: MFSK Constellation diagram for (a)M=2 and (b)M=3.

Constellation diagram of MFSK is shown in Fig. 2.11. The ith signal point is located on the ith corresponding axis $\varphi_i(t)$ at a displacement of $\sqrt{E_s}$ from the origin of the signal space. In this modulation scheme the distance between any two signal points $s_i(t)$ and $s_k(t)$ is constant

$$d_{i,k} = ||\mathbf{s_i} - \mathbf{s_k}|| = \sqrt{2E_s} = d_{min} \quad ; i \neq k \tag{2.101}$$

The demodulator for coherent MFSK is exactly the one shown in Fig. 2.4. As all signals are equidistant from each other, to derive the average probability of symbol error, we may assume that signal $s_0(t)$ is sent without losing generality. The received signal in this case is $r(t) = s_0(t) + n(t)$ where $n(t)$ denotes a Gaussian process with zero mean and variance $N_0/2$. The M observable variables $\{r_j\}_{j=0}^{M-1}$ at the output of M correlators are

$$r_0 = \sqrt{E_s} + n_0 \quad ; j = 0 \tag{2.102a}$$

$$r_j = n_j \qquad ; j = 1, 2, \ldots, M-1 \tag{2.102b}$$

where $\{n_j\}_{j=0}^{M-1} \sim \mathcal{N}(0, N_0/2)$ and is defined by

$$n_j = \int_{kT_s}^{(k+1)T_s} n(t)\varphi_j(t)dt \quad ; j \in \{0, 1, \ldots, M-1\} \tag{2.103}$$

For a given r_0, the probability that each of these $M-1$ variables $r_1, r_2, \ldots r_{M-1}$ are less than r_0 is

$$
\begin{aligned}
\Pr\left[r_j < r_0 | r_0 \atop \forall j \neq 0\right] &= \Pr[(r_1 \leq r_0) \cap \ldots \cap (r_{M-1} \leq r_0) | r_0] \\
&= \{\Pr[r_j \leq r_0 | r_0]\}^{M-1} \quad ; \forall j \neq 0 \\
&= \left\{\frac{1}{\sqrt{\pi N_0}} \int_{-\infty}^{r_0} \exp\left(-\frac{u^2}{N_0}\right) du\right\}^{M-1}
\end{aligned}
\tag{2.104}
$$

assuming each of the $r_j; \forall j \neq 0$ are independent random variables. The probability that the demodulator will make an incorrect decision is simply

$$
\begin{aligned}
P_e &= 1 - \int_{-\infty}^{\infty} \Pr\left[r_j < r_0 | r_0 \atop \forall j \neq 0\right] f_{r_0}(r_0) dr_0 \\
&= 1 - \frac{1}{\sqrt{\pi N_0}} \int_{-\infty}^{\infty} \exp\left[-\frac{(r_0 - \sqrt{E_s})^2}{N_0}\right] \\
&\quad \times \left\{\frac{1}{\sqrt{\pi N_0}} \int_{-\infty}^{r_0} \exp\left(-\frac{u^2}{N_0}\right) du\right\}^{M-1} dr_0
\end{aligned}
\tag{2.105}
$$

Using the definition of Gaussian Q function $Q(z) = 1/\sqrt{2\pi} \int_z^\infty \exp(-u^2/2)du$ and the property $Q(-z) = 1 - Q(z)$, the average symbol error probability is given as

$$
\begin{aligned}
P_e = 1 - \frac{1}{\sqrt{\pi N_0}} \int_{-\infty}^{\infty} &\exp\left[-\frac{(r_0 - \sqrt{E_s})^2}{N_0}\right] \\
&\times \left[1 - Q\left(r_0\sqrt{\frac{2}{N_0}}\right)\right]^{M-1} dr_0
\end{aligned}
\tag{2.106}
$$

Finally, with the substitution $x = r_0\sqrt{2/N_0}$ the exact SEP expression for equal-energy, equiprobable, orthogonal signal set with coherent detection in AWGN channel is given by [2, 6]

$$P_e = 1 - \frac{1}{\sqrt{2\pi}} \int_{-\infty}^{\infty} \exp\left[-\frac{(x - \sqrt{2\eta})^2}{2}\right] [1 - Q(x)]^{M-1} dx \tag{2.107}$$

An infinite series containing infinite range integrals may be obtained by expanding the term $[1 - Q(x)]^{M-1}$ in (2.107) using binomial theorem, out of which first 6 terms are [50]

$$
\begin{aligned}
P_e \approx\; & 1 - \frac{1}{\sqrt{2\pi}} \int_{-\infty}^{\infty} f(x,\eta)dx \\
& + \frac{M-1}{\sqrt{2\pi}} \int_{-\infty}^{\infty} f(x,\eta)Q(x)dx \\
& - \frac{(M-1)(M-2)}{2\sqrt{2\pi}} \int_{-\infty}^{\infty} f(x,\eta)Q^2(x)dx \\
& + \frac{(M-1)(M-2)(M-3)}{6\sqrt{2\pi}} \int_{-\infty}^{\infty} f(x,\eta)Q^3(x)dx \\
& - \frac{(M-1)(M-2)(M-3)(M-4)}{24\sqrt{2\pi}} \int_{-\infty}^{\infty} f(x,\eta)Q^4(x)dx \\
& \;\;\ldots\ldots\ldots
\end{aligned}
\tag{2.108}
$$

where $f(x,\eta) = \exp[-(x - \sqrt{2\eta})^2/2]$. This truncation was inspired by the fact that we can have alternate representations up to the fourth power of Q function [11]. The second term in (2.108) merely consists of Gaussian PDF resulting unity and thus cancels out the first term. Also using the identity $1/\sqrt{2\pi} \int_{-\infty}^{\infty} Q(x)\exp[-(x-\sigma)^2/2]dx = Q(\sigma/\sqrt{2})$ [51] and alternate representation of Q function [10], the third term may be expressed with an integral containing exponential function only. For the rest of the terms, alternate expressions of $Q^n(x)$ in terms of finite range integral may be obtained from Simon [11]. With a change of order of integration and using the following result

$$
\begin{aligned}
& \frac{1}{\sqrt{2\pi}} \int_{-\infty}^{\infty} \exp\left[-\frac{(x-\sqrt{2\eta})^2}{2}\right] \exp\left(-\frac{x^2}{2\sin^2\theta}\right) dx \\
& = \frac{\sin\theta}{\sqrt{1 + \sin^2\theta}} \exp\left(-\frac{\eta}{1 + \sin^2\theta}\right)
\end{aligned}
\tag{2.109}
$$

(2.108) can be written as

$$
\begin{aligned}
P_e =\; & \frac{M-1}{\pi} \int_0^{\pi/2} \exp\left(-\frac{\eta}{2\sin^2\theta}\right) d\theta \\
& - \frac{(M-1)(M-2)}{2\pi} \int_0^{\pi/4} f_1(\theta,\eta)d\theta \\
& + \frac{(M-1)(M-2)(M-3)}{6\pi^2} \int_0^{\pi/6} f_1(\theta,\eta)f_2(\theta)d\theta \\
& + \frac{(M-1)(M-2)(M-3)}{12\pi^2} \int_0^{\sin^{-1}\left(\frac{1}{\sqrt{3}}\right)} f_1(\theta,\eta)[\pi - f_2(\theta)]d\theta \\
& - \frac{(M-1)(M-2)(M-3)(M-4)}{24\pi^2} \int_0^{\pi/6} f_1(\theta,\eta)f_2(\theta)d\theta
\end{aligned}
\tag{2.110}
$$

37

where

$$f_1(\theta, \eta) = \frac{\sin \theta}{\sqrt{1 + \sin^2 \theta}} \exp\left(-\frac{\eta}{1 + \sin^2 \theta}\right) \qquad (2.111)$$

and

$$f_2(\theta) = \cos^{-1}\left(\frac{3\cos 2\theta - 1}{2\cos^3 2\theta} - 1\right) \qquad (2.112)$$

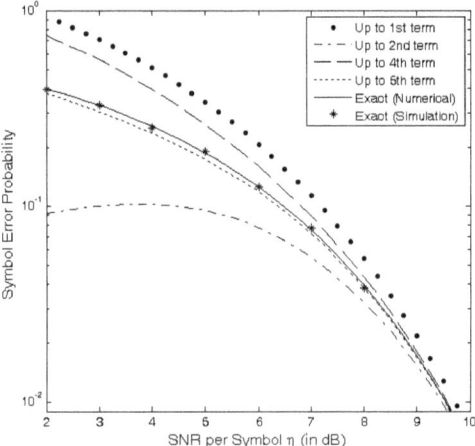

Figure 2.12: Comparison of the proposed approximation for SEP of coherent MFSK ($M = 10$) with exact values (numerical and simulation) and earlier bound [2] given by the first term in (2.110).

The expression in (2.110) gives exact result up to $M = 5$ and for higher values of M it offers a good approximation. For fixed M these bounds become increasingly tight as SNR per symbol η is increased. For completeness, we would like to mention that, neglecting all the terms in (2.110) except the first, we arrive at

$$P_e \leq \frac{M-1}{2}\text{erfc}\left(\sqrt{\frac{\eta}{2}}\right) = \frac{M-1}{\pi}\int_0^{\pi/2} \exp\left(-\frac{\eta}{2\sin^2 \theta}\right) d\theta \qquad (2.113)$$

the standard available SEP approximation for coherent MFSK [2]. In Fig. 2.12, a comparison of the proposed approximation with the existing upper bound [2] is made. The exact values are also calculated through direct numerical integration while simulation results confirm those values. The figure clearly shows that the gap between exact and approximated values diminishes as more number of terms in (2.110) are considered. When all the terms are taken into account, the values obtained from (2.110) are far superior than to the loose upper bound [2]. A tighter bound was presented by Hughes [51, (1.2)] later on

$$P_e \leq 1 - \left[1 - \frac{1}{2}\text{erfc}\left(\sqrt{\frac{\eta}{2}} \right) \right]^{M-1}$$

(2.114)

To derive the average BEP we note that all symbol errors occur with equal probability

$$\frac{P_e}{M-1} = \frac{P_e}{2^n - 1}$$

(2.115)

Since there are $\binom{n}{k}$ ways in which k out of n bits in a symbol may be in error, the average number of bit errors per n bit symbol is

$$\sum_{k=1}^{n} k \binom{n}{k} \frac{P_e}{2^n - 1} = n \frac{2^{n-1}}{2^n - 1} P_e$$

(2.116)

Because there are n bits in a symbol, the average BEP is simply

$$P_{e,b} = \frac{2^{n-1}}{2^n - 1} P_e \approx \frac{P_e}{2} \quad ; n \gg 1$$

(2.117)

Utilizing the relation between BEP and SEP of coherent MFSK an approximated BEP can be calculated by substituting (2.110) in (2.117). It is interesting to note that in AWGN channel, for a given E_s/N_0 SEP of MFSK increases with M, but for a given E_b/N_0 the BEP decreases as M increases.

2.3.5 Non-coherent MFSK

Orthogonal modulation with non-coherent detection is a practical choice for situations where the received signal phase cannot be reliably estimated. Important examples include military communications using fast frequency hopping, airborne communications with high Doppler shifts due to significant relative motion of the transmitter and receiver, and high phase noise scenarios due to the use of inexpensive or unreliable local oscillators. Widespread use of non-coherent MFSK (ncMFSK), absence of costly and complex phase synchronized receivers, and analytical intractability of probability of error expressions for coherent MFSK inspired a myriad of research papers [52, 53, 54, 55, 56, 57, 58] on the performance of the non-coherent version of MFSK in wireless fading channels.

Let us assume, without losing generality, that the transmitter sends the 0th message a_0 as governed by the current input bit pattern to the modulator. Accordingly signal $s_0(t)$ would be produced and fed to the channel. The corresponding received signal is $r(t) = s_0(t) + n(t)$, where $n(t)$ accounts for the noise component. Under the AWGN channel assumption, $n(t)$ refers to a white noise process which is Gaussian distributed with zero mean and variance $N_0/2$. Fig. 2.13 shows a demodulator structure for ncMFSK

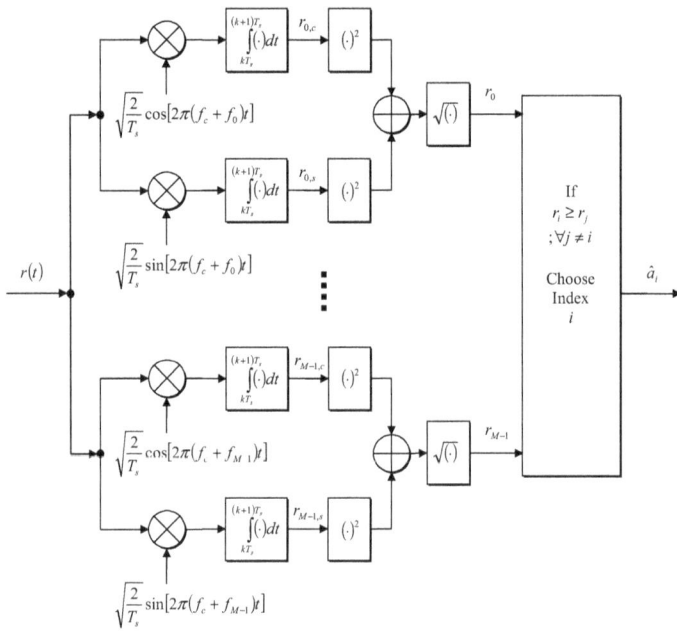

Figure 2.13: Matched filter equivalent demodulator for ncMFSK.

which is equivalent to the structure shown in Fig. 2.5. Referring to Fig. 2.13 one may easily find that the output of the product integrators $\{r_{j,c}, r_{j,s}\}_{j=0}^{M-1}$ are

$$r_{0,c} = \sqrt{E_s} + n_{0,c} \quad ; j = 0 \tag{2.118a}$$

$$r_{j,c} = n_{j,c} \quad ; j = 1, 2, \ldots, M-1 \tag{2.118b}$$

$$r_{j,s} = n_{j,s} \quad ; j = 0, 1, \ldots, M-1 \tag{2.118c}$$

where $\{n_{j,c}, n_{j,s}\}_{j=0}^{M-1} \sim \mathcal{N}(0, N_0/2)$ and are defined by

$$n_{j,c} = \int_{kT_s}^{(k+1)T_s} n(t)\sqrt{\frac{2}{T_s}}\cos[2\pi(f_c + f_j)t]dt \qquad ;\forall j \tag{2.119a}$$

$$n_{j,s} = \int_{kT_s}^{(k+1)T_s} n(t)\sqrt{\frac{2}{T_s}}\sin[2\pi(f_c + f_j)t]dt \qquad ;\forall j \tag{2.119b}$$

From (2.118) it turns out that the decision variable $r_j = \sqrt{r_{j,c}^2 + r_{j,s}^2}$ will be Rayleigh distributed for $j = 1, 2, \ldots, M-1$ [18]. The corresponding PDF and cumulative distribution function (CDF) are given by

$$f_{r_j}(u) = \frac{2u}{N_0}\exp\left(-\frac{u^2}{N_0}\right) \qquad ;u \geq 0, j \neq 0 \tag{2.120}$$

$$F_{r_j}(u) = 1 - \exp\left(-\frac{u^2}{N_0}\right) \qquad ;u \geq 0, j \neq 0 \tag{2.121}$$

However the decision variable $r_0 = \sqrt{r_{0,c}^2 + r_{0,s}^2}$ will be Rician distributed [18] having a PDF

$$f_{r_0}(u) = \frac{2u}{N_0}\exp\left(-\frac{u^2 + E_s}{N_o}\right) I_0\left(u\frac{2\sqrt{E_s}}{N_0}\right) \qquad ;u \geq 0 \tag{2.122}$$

where $I_0(\cdot)$ is modified Bessel function of zero order and of first kind [9, (9.6.16)]. In terms of the SNR per symbol $\eta = E_s/N_0$ the PDF can also be written as

$$f_{r_0}(u) = \frac{2\exp(-\eta)}{N_0}u\exp\left(-\frac{u^2}{N_o}\right) I_0\left(u\sqrt{\frac{4\eta}{N_0}}\right) \qquad ;u \geq 0 \tag{2.123}$$

For a given r_0, the probability that each of the $M-1$ variables $r_1, r_2, \ldots r_{M-1}$ are less than r_0 is

$$
\begin{aligned}
\Pr\left[\underset{\forall j \neq 0}{r_j < r_0|r_0}\right] &= \Pr[(r_1 \leq r_0) \cap \ldots \cap (r_{M-1} \leq r_0)|r_0] \\
&= \{F_{r_j}(r_0)\}^{M-1} \quad ;\forall j \neq 0 \\
&= \left\{1 - \exp\left(-\frac{r_0^2}{N_0}\right)\right\}^{M-1} \\
&= \sum_{k=0}^{M-1}(-1)^k\binom{M-1}{k}\exp\left(-\frac{kr_0^2}{N_0}\right)
\end{aligned}
\tag{2.124}
$$

using simple binomial expansion $(1-x)^n = \sum_{k=0}^{n}\binom{n}{k}(-x)^k$ and assuming that each of the

$r_j; \forall j \neq 0$ are independent random variables. Thus the probability that the demodulator will make an incorrect decision is

$$
\begin{aligned}
P_e &= 1 - \int_0^\infty \Pr\left[\underset{\forall j \neq 0}{r_j < r_0 | r_0}\right] f_{r_0}(r_0) dr_0 \\
&= 1 - \sum_{k=0}^{M-1} (-1)^k \binom{M-1}{k} \int_0^\infty \exp\left(-\frac{kr_0^2}{N_0}\right) f_{r_0}(r_0) dr_0 \\
&= \sum_{k=1}^{M-1} (-1)^{k+1} \binom{M-1}{k} \int_0^\infty \exp\left(-\frac{kr_0^2}{N_0}\right) f_{r_0}(r_0) dr_0
\end{aligned}
\tag{2.125}
$$

Putting the value of $f_{r_0}(r_0)$ from (2.123) in (2.125) results in

$$
\begin{aligned}
P_e &= \frac{2\exp(-\eta)}{N_0} \sum_{k=1}^{M-1} (-1)^{k+1} \binom{M-1}{k} \\
&\times \int_0^\infty r_0 \exp\left[-\frac{(k+1)r_0^2}{N_0}\right] I_0\left(r_0\sqrt{\frac{4\eta}{N_0}}\right) dr_0
\end{aligned}
\tag{2.126}
$$

Further substituting $x = r_0\sqrt{2(k+1)/N_0}$ and $y = \sqrt{2\eta/(k+1)}$ in the integral given in (2.126) we get

$$
\begin{aligned}
P_e &= \sum_{k=1}^{M-1} \frac{(-1)^{k+1}}{k+1} \binom{M-1}{k} \exp\left(-\frac{k\eta}{k+1}\right) \\
&\times \int_0^\infty x \exp\left(-\frac{x^2+y^2}{2}\right) I_0(xy) dx
\end{aligned}
\tag{2.127}
$$

Noting that $\int_0^\infty x\exp[-(x^2+y^2)/2]I_0(xy)dx = 1$, where the integrand is just a Rician PDF [18], we may finally write the symbol error probability of ncMFSK over AWGN channel as

$$
P_e = \sum_{k=1}^{M-1} \frac{(-1)^{k+1}}{k+1} \binom{M-1}{k} \exp\left(-\frac{k\eta}{k+1}\right)
\tag{2.128}
$$

or alternatively,

$$
P_e = \frac{1}{M} \sum_{k=2}^{M} (-1)^k \binom{M}{k} \exp\left[-\eta\left(1-\frac{1}{k}\right)\right]
\tag{2.129}
$$

On expanding the sum in (2.128) and considering the first term only a simple upper bound for ncMFSK may be found [59]

$$
P_e \leq \frac{M-1}{2} \exp\left(-\frac{\eta}{2}\right)
\tag{2.130}
$$

Keeping the symbol length T_s constant, the information rate may be increased by using an increasing number of tones or, the M value. In this case the bound in (2.130) indicates that SNR (η) needs only to be increased by an order of $\ln(M)$ to keep the symbol error rate constant. Further, while the SEP is given by (2.128) or (2.129), the average BEP for ncMFSK may be found by substituting (2.117) in (2.128) or (2.129).

2.3.6   MQAM

Quadrature amplitude modulation belongs to the class of non-constant envelope modulation and is suitable for bandwidth efficient communication systems. M-ary quadrature amplitude modulation (MQAM) has two degrees of freedom, i.e. the information bits are encoded in both the amplitude and phase of the transmitted signal. As a result, MQAM is more spectrally-efficient than M-ary pulse amplitude modulation (MPAM) and MPSK, in that it can encode the most number of bits per symbol for a given average energy [4]. MQAM is currently being employed in many high data rate wireless applications including digital audio broadcasting (DAB) and terrestrial, satellite, cable or handheld digital video broadcasting (DVB-T/S/C/H). QAM is also widely used in modems designed for telephone channels. The international telecommunication union (ITU) telephone circuit modem standards V.29 to V.33 are all based on various QAM schemes ranging from uncoded 16-QAM to trellis coded 128-QAM. The research of QAM applications in satellite systems, point-to-point wireless systems, and mobile cellular telephone systems also has been very active [60].

The square MQAM constellation as shown in Fig. 2.14 was first proposed by Campopiano and Glazer in 1962 [61]. Research results [60] have shown that the square constellation is the most appropriate choice in AWGN channels. A few of the other constellations (star, rectangular, circular etc.) offer slightly better error performance, but with a much more complicated system implementation. Therefore we will concentrate solely on the square constellation in this book.

A general description of MQAM signal set $\{s_i(t)\}_{i=0}^{M-1}$ may be obtained by setting $A_i = \sqrt{2E_i/T_s}$ and $f_i = 0$ in (2.15)

$$s_i(t) = \sqrt{\frac{2E_i}{T_s}} \cos(2\pi f_c t + \phi_i) \quad ; 0 \leq t \leq T_s \tag{2.131}$$

where E_i denotes energy content of the ith symbol waveform. Expression (2.131) can be written in the quadrature form as

$$s_i(t) = \sqrt{\frac{2E_i}{T_s}} \cos(\phi_i) \cos(2\pi f_c t) - \sqrt{\frac{2E_i}{T_s}} \sin(\phi_i) \sin(2\pi f_c t) \tag{2.132}$$

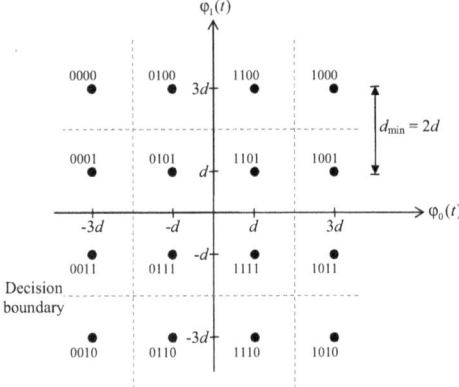

Figure 2.14: Square MQAM ($M = 16$) constellation diagram with Gray coding.

Also, as evident from (2.24), the equivalent baseband signal of (2.131) would be

$$\tilde{s}_i(t) = \sqrt{\frac{2E_i}{T_s}} \exp(j\phi_i) \quad ; 0 \le t \le T_s \tag{2.133}$$

The MQAM signal set can be expressed as a linear combination of two orthonormal basis functions

$$\varphi_0(t) = \sqrt{\frac{2}{T_s}} \cos(2\pi f_c t) \tag{2.134a}$$

$$\varphi_1(t) = \sqrt{\frac{2}{T_s}} \sin(2\pi f_c t) \tag{2.134b}$$

where the coordinates of the signal constellation points would be

$$s_{i0} = \int_0^{T_s} s_i(t)\varphi_0(t)dt = \sqrt{\frac{2E_i}{T_s}} \cos(\phi_i) \tag{2.135a}$$

$$s_{i1} = \int_0^{T_s} s_i(t)\varphi_1(t)dt = \sqrt{\frac{2E_i}{T_s}} \sin(\phi_i) \tag{2.135b}$$

Further, from (2.33), the distance between any two MQAM signal constellation points s_i and s_k is given by

$$d_{ik} = ||s_i - s_k|| = \sqrt{(s_{i0} - s_{k0})^2 + (s_{i1} - s_{k1})^2} \tag{2.136}$$

Square MQAM ($M = P^2; P = 2^n, n = 1, 2, 3, \dots$) signal waveforms consist of two

44

independent MPAM carriers in quadrature. In this case equation (2.132) takes the form

$$s_i(t) = \sqrt{\frac{2E_0}{T_s}} u_i \cos(2\pi f_c t) - \sqrt{\frac{2E_0}{T_s}} v_i \sin(2\pi f_c t) \qquad (2.137)$$

where (u_i, v_i) is an element of the $P \times P$ matrix

$$\{u_i, v_i\} = \begin{bmatrix} (-P+1, P-1) & (-P+3, P-1) & \cdots & (P-1, P-1) \\ (-P+1, P-3) & (-P+3, P-3) & \cdots & (P-1, P-3) \\ \vdots & \vdots & & \vdots \\ (-P+1, -P+1) & (-P+3, -P+1) & \cdots & (P-1, -P+1) \end{bmatrix} \qquad (2.138)$$

and the pair $(u_i d, v_i d) = (s_{i0}, s_{i1})$ denotes the ith message point $\mathbf{s_i}$ in the two dimensional signal space spanned by $\varphi_0(t)$ and $\varphi_1(t)$. The minimum distance (d_{min}) between two message points in the constellation is given by $d_{min} = 2d$, where $d = \sqrt{E_0}$ and E_0 is the energy of the signal with the lowest amplitude.

The average signal energy per symbol may be calculated as

$$
\begin{aligned}
E_s &= \frac{d^2}{M} \sum_{i=0}^{M-1} (u_i^2 + v_i^2) = \frac{2E_0}{P} \sum_{i=0}^{P-1} u_i^2 \\
&= \frac{4E_0}{P} \sum_{i=0}^{P/2-1} (2i+1)^2 \\
&= \frac{2(P^2-1)E_0}{3} = \frac{2(M-1)E_0}{3}
\end{aligned}
\qquad (2.139)
$$

The first multiplying factor 2 comes due to equal contributions are made by the in-phase and quadrature components while the second 2 factor accounts for the symmetric nature of the pertinent amplitude levels around the origin of signal space. We have also assumed that the P amplitude levels of the in-phase and quadrature components are equally likely which enables us to compute the average over P terms instead of M. Further the identity $\sum_{k=0}^{n}(2k+1)^2 = (n+1)(2n+1)(2n+3)/6$, which may be easily deduced from the formula of sum of the squares of the first n natural numbers $\sum_{k=0}^{n} k^2 = n(n+1)(2n+1)/6$, was used to derive (2.139).

Fig. 2.15 shows a coherent MQAM demodulator structure based on ML criterion. The demodulator consists of two correlator blocks whose outputs are fed to a decision device. The device decides in favour of a signal which is closest to the received signal in the vector space. Since the in-phase and quadrature components of MQAM form two independent PAM signal with $(P = \sqrt{M})$ amplitude levels, the probability of correct detection for such a scheme may be written as

$$P_c = (1 - P_{e(PAM)})^2 \qquad (2.140)$$

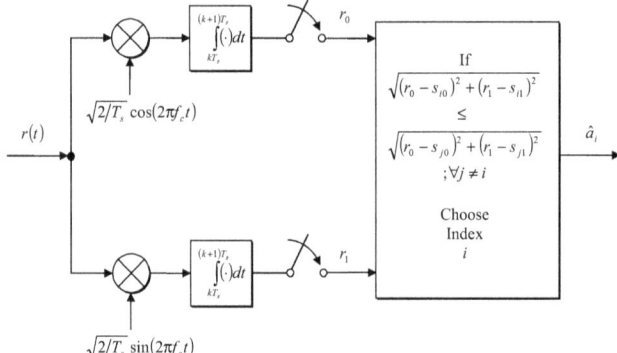

Figure 2.15: Coherent square MQAM demodulator using two correlators.

where $P_{e(MPAM)}$ is the probability of error for either component in AWGN channel. Naturally the SEP for MQAM would be then

$$P_e = 1 - (1 - P_{e(PAM)})^2 \tag{2.141}$$

Evaluation of SEP for MPAM is rather straightforward. Assuming $s_i(t)$ is transmitted, i.e. the received signal being $r(t) = s_i(t) + n(t)$, a symbol error occurs when the noise $n(t)$ exceeds in magnitude one-half of the distance between two adjacent levels. This probability is same for each $s_i(t)$ except for the two outside levels, where an error can occur in one direction only. Assuming all amplitude levels are equally likely, the average SEP is

$$P_{e(PAM)} = \frac{P-1}{P} \Pr[|\mathbf{r} - \mathbf{s_i}| > d] \tag{2.142}$$

Thus from (2.6)

$$
\begin{aligned}
P_{e(PAM)} &= \frac{P-1}{P} \frac{2}{\sqrt{\pi N_0}} \int_d^\infty \exp\left(-\frac{x^2}{N_0}\right) dx \\
&= \frac{P-1}{P} \mathrm{erfc}\left(\sqrt{\frac{E_0}{N_0}}\right)
\end{aligned}
\tag{2.143}
$$

Substituting (2.143) in (2.141) we get

$$P_e = \frac{2(P-1)}{P} \mathrm{erfc}\left(\sqrt{\frac{E_0}{N_0}}\right) - \left(\frac{P-1}{P}\right)^2 \mathrm{erfc}^2\left(\sqrt{\frac{E_0}{N_0}}\right) \tag{2.144}$$

and in terms of the average symbol energy $E_s = 2E_0(M-1)/3$ it becomes [62, 63]

$$P_e = 2\left(1 - \frac{1}{\sqrt{M}}\right) \text{erfc}\left[\sqrt{\frac{3\eta}{2(M-1)}}\right]$$
$$- \left(1 - \frac{1}{\sqrt{M}}\right)^2 \text{erfc}^2\left[\sqrt{\frac{3\eta}{2(M-1)}}\right] \tag{2.145}$$

It is interesting to note that SEP performance of 4QAM and QPSK are identical, which can be verified by noting the similarity between (2.74) and (2.145) for $M = 4$. The SEP given by (2.145) is sometimes written in terms of average SNR per bit η_b

$$P_e = 2q\text{erfc}(\sqrt{p\eta_b}) - q^2\text{erfc}^2(\sqrt{p\eta_b}) \tag{2.146}$$

where $p = (1.5)\log_2 M/(M-1)$ and $q = 1 - 1/\sqrt{M}$ [64, 65, 66, 67, 68]. Also, using alternate expressions of $\text{erfc}(\cdot)$ and $\text{erfc}^2(\cdot)$, equation (2.146) may be written as

$$P_e = \frac{4q}{\pi}\left[\int_0^{\frac{\pi}{2}} \exp\left(-\frac{p\eta_b}{\sin^2\theta}\right) d\theta - q\int_0^{\frac{\pi}{4}} \exp\left(-\frac{p\eta_b}{\sin^2\theta}\right) d\theta\right] \tag{2.147}$$

which is useful in MGF based computations [69]. An approximated BEP for Gray coded MQAM may be found by substituting (2.58) in (2.146). However, finding actual BEP is in general very tedious and no general formula in terms of the constellation size is known. A detailed coverage on various BEP approximation for MQAM may be found in Hanzo's book[60].

2.4 Comparison of Modulation Schemes

This concluding section incorporates a comparative study of the modulation schemes discussed so far. All the modulation methods are first compared on the basis of SEP and then on the basis of BEP for a given value of constellation size M. Fig. 2.16 portrays such SEP performances for $M = 16$, while Fig. 2.17 gives the comparative BEP performance for the same M value. The abscissa for the former plot is SNR per symbol (η) whereas for the later SNR per bit (η_b) has been used.

From both the figures it is evident that the error performance of the PSK family (MPSK and MDPSK) is poorer compared to the FSK family (MFSK and ncMFSK) in general. Performance of MQAM is better than pure PSK but FSK outperforms QAM too. Also the coherent demodulation methods (MFSK and MPSK) yield better results than their non-coherent (ncMFSK) or differentially coherent (MDPSK) counterparts. A horizontal movement (keeping the P_e or $P_{e,b}$ fixed) from left to right in the graph tells how the SNR required to maintain a fixed error rate increases from FSK to PSK or from

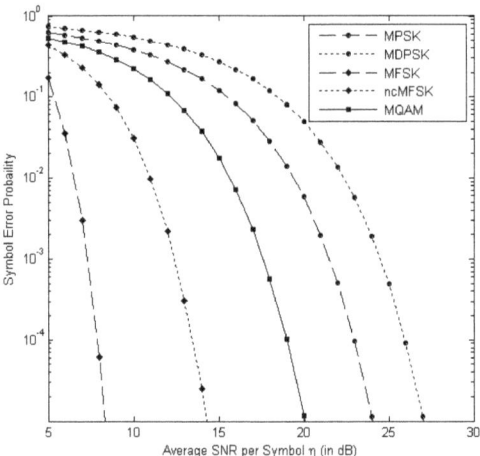

Figure 2.16: SEP comparison of different M-ary ($M = 16$) modulations.

PSK to DPSK. Similarly a vertical movement (keeping SNR η or η_b fixed) from top to bottom displays how the error rates fall when we switch to MPSK from MDPSK etc.

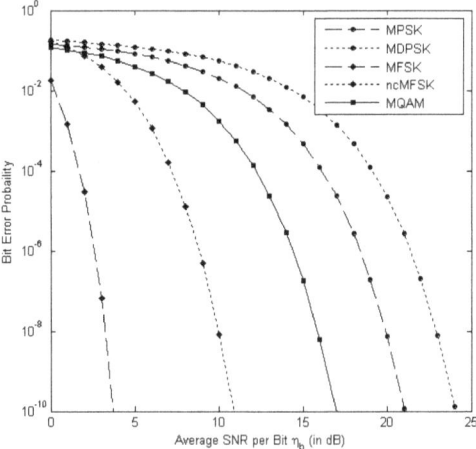

Figure 2.17: BEP comparison of different M-ary ($M = 16$) modulations.

However the comparison is incomplete unless we introduce the other most important parameter, transmission bandwidth. Fig. 2.16 and Fig. 2.17 shows only the trade-off between SNR and error probability, and thus may lead to inappropriate conclusions. Thus, in Fig. 2.18, we also compare the trade-off between SNR and bandwidth of different modulations, keeping the error rate fixed ($P_{e,b} = 10^{-2}$).

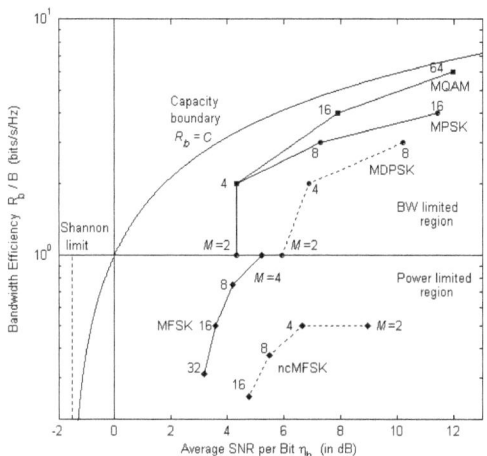

Figure 2.18: Bandwidth efficiency - power efficiency tradeoff of different M-ary modulations for a given BEP ($P_{e,b} = 10^{-2}$).

The ordinate for the figure is taken as the normalized channel bandwidth R_b/B, where R_b is the bit rate (in bits/s) and B is the bandwidth (in Hz). Thus the parameter R_b/B, denoting how many bits per second may be transmitted per unit channel bandwidth, has a unit of bits/s/Hz and thereby serves as a measure of how efficiently bandwidth is utilized. The upper limit of bandwidth efficiency is governed by the capacity boundary which may be derived from Shannon Hartley law and Shannon's channel coding theorem. In fact, from the Shannon Hartley law we have the following relation $C = B \log_2(1 + \eta_b R_b/B)$. Now with no room for channel coding ($R_b = C$) we obtain the following boundary

$$\eta_b = \frac{2^{R_b/B} - 1}{R_b/B} \tag{2.148}$$

which gives the upper limit for reliable communication as demonstrated by Shannon's channel coding theorem. The abscissa for Fig. 2.18 is SNR per bit (η_b) giving a measure of power efficiency. The capacity limit shows that SNR may be reduced (as long it stays above the Shannon's limit of -1.6 dB) at the expense of bandwidth. Fig. 2.18 is often referred to as the bandwidth efficiency - power efficiency diagram.

For plotting the curves, first the BEP value ($P_{e,b} = 10^{-2}$) is transformed into the equivalent SEP value with the help of the SEP-BEP relation defined for each modulation family in Section 2.3. From the SEP expressions, as available in Section 2.3, the SNR (η) requirement for the given SEP is then calculated. Finally the SNR per bit (η_b) is calculated by the formula $\eta_b = \eta / \log_2 M$. When the SEP expressions are not available in closed form, tight bounds have been used instead. The procedure for calculating the η_b value is repeated for each modulation order M.

Every point in the diagram is characterized by the pair $\{R_b/B, \eta_b\}$. Thus bandwidth efficiency for each M value for a given family of modulation also needs to be computed. The calculation of bandwidth efficiency for MPSK, MDPSK, and MQAM is rather simple when we assume a bandwidth $B = 1/T_s = R_s$ is required to send a symbol of duration T_s. The corresponding bandwidth efficiency per bit is thus

$$\frac{R_b}{B} = \frac{R_s \log_2 M}{B} = \log_2 M \tag{2.149}$$

Quite naturally the efficiency increases as M increases. Orthogonal signals, on the other hand, have totally different bandwidth requirements. For coherent MFSK we require M orthogonal carriers with a minimum frequency separation of $1/2T_s$ to maintain orthogonality [2]. Thus the bandwidth efficiency in this case is

$$\frac{R_b}{B} = \frac{R_s \log_2 M}{M/2T_s} = \frac{2 \log_2 M}{M} \tag{2.150}$$

For ncMFSK the transmitter require a wider separation of $1/T_s$ between carriers [6] and thus the efficiency for the non-coherent case is

$$\frac{R_b}{B} = \frac{R_s \log_2 M}{M/T_s} = \frac{\log_2 M}{M} \tag{2.151}$$

As $\log_2 M$ increases at a rate quite lower than M, from (2.150) and (2.151) one may find that the bandwidth efficiency for FSK family decreases with M.

In summary, for MPSK, MDPSK and MQAM, increasing M results in a higher bandwidth efficiency R_b/B value. However, the cost of achieving higher data rate is an increase in the SNR per bit (η_b). Consequently, these modulation methods are appropriate for communication channels that are bandwidth limited ($R_b/B > 1$) but the transmitter may support sufficiently high η_b to support the increase in M. Between the three modulation schemes, MQAM offers better trade-off than PSK, but often requires linear amplifiers which limit the ability to increase the transmitter power beyond a certain level. MDPSK, on the other hand comes with the advantage of simpler receiver structure.

M-ary orthogonal signalling yield a bandwidth efficiency value ($R_b/B \leq 1$). As M increases R_b/B decreases due to an increase in the required channel bandwidth. However, the SNR per bit (η_b) required to achieve a given BEP decreases as M increases. Consequently M-ary orthogonal signals are appropriate for power-limited channels that have sufficiently large bandwidth to accommodate a large number of signals.

The trade-offs associated with the bandwidth limited region and the power limited region in Fig. 2.18 is not equitable [59]. For the bandwidth limited region, the capacity boundary curve is flattened, and achieving a larger R_b/B requires an exponential rise in η_b. In the power limited region, however, the capacity boundary curve is steep. This means for a small reduction in required η_b a large reduction in R_b/B is warranted. Further the choice of modulation scheme described here is applicable to uncoded system. For systems employing error correction coding, modulation selection is not trivial as a different power bandwidth trade-off may exist.

Chapter Summary

Digital modulation refers to the process of converting binary bits into signal waveforms suitable for transmission through the physical communication medium. This book examines error performance of digital modulated signals over wireless fading channels. Thus, a thorough description of these modulation schemes and their performance in absence of fading are prerequisites for developing the error expressions over fading channels. In this chapter, we have presented BEP and SEP expressions in non-fading AWGN channel for the modulations considered in the rest of the book.

Chapter 3

Fading, Diversity and Combining

Future generation wireless systems are envisioned to provide highly mobile, ever present, broadband communication, that too over heterogeneous radio interfaces. These conflicting demands are often difficult to support due to three major constraints in the system design: limited *transmitter power*, scarce *radio bandwidth* and a complex and harsh *time-varying radio channel*. As transmit power and bandwidth are fundamental limitations of every communication system and well studied, the current chapter is focused at the random wireless channel that needs some special attention.

The major goal of this chapter is to lay a mathematical framework for statistical modelling of wireless channels which will be used in the subsequent chapters for performance analysis over different wireless channels. We devote the first two sections, Section 3.1 and Section 3.2 respectively, to provide brief overviews on fading mechanisms and different fading models. This is followed by a discussion on diversity, the predominant fading mitigation technique used in practice, in Section 3.3. Finally, Section 3.4 describes different combining schemes used to process the diversity branches.

3.1 Fading Mechanisms

3.1.1 What is Fading?

For radio waves reflection, diffraction and scattering are three main propagation mechanisms apart from direct line of sight (LOS) transmission [70, 71]. *Reflection* occurs when signal wave is obstructed by a smooth surface with very large dimensions compared to signal wavelength (e.g. Earth, buildings, walls) and may cause destructive interference at the receiver. *Diffraction*, on the other hand, occurs when signal is obstructed by a surface having sharp irregularities or edges with dimensions comparable to the signal wavelength. Diffraction causes secondary waves to be formed behind the obstructing body. *Scattering* occurs when the signal travels through a medium consisting of objects with dimensions that are on the order or less of the wavelength (e.g. lamp posts, street signs, foliage), and where the number of such obstacles per unit volume is large. Due to scattering the reflected energy spreads out in all directions.

In wireless channels multifold reflection, diffraction and scattering result in unwanted variation of the received signal. Typically, the received signal is a sum of the components arising from the above three phenomena. The strength of the received signal fluctuates rapidly with respect to time and the displacement of the transmitter and the receiver. All these type of distortions are collectively termed as fading.

3.1.2 Large-scale and Small-scale Fading

In a communication system fading refers to random attenuation of received signal power. The variation in received signal power can be characterized with path loss, shadowing and multipath effects. *Path loss* is caused by dissipation of the power radiated by the transmitter over distance. *Shadowing* is caused by obstacles between the transmitter and receiver. *Multipath* effects are observed due to random scatterers present in the propagation medium.

Variation due to path loss occurs over very large distances (100 m - a few km). For shadowing, the variation is over a distance proportional to the length of the obstructing objects (10 m - 100 m). Since variations due to path loss and shadowing occur over relatively large distances, this variation is referred to as *large-scale fading*. Variation due to multipath occurs over very short distances, on the order of the signal wavelength ($<$ 1 m). As multipath causes variation over very small distances, this variation is referred to as *small-scale fading*.

Large scale effects are due to general terrain, density and height of buildings, vegetation etc., and the effects have a behavior that varies slowly with time. Further, large-scale fading is important for predicting the coverage and availability of a particular service. Compared to this, small scale effects are due to local environment, nearby trees, buildings etc., and these effects have a much shorter time scale. A consideration of small scale effects is important for the design of the modulation format and for general transmitter and receiver design.

3.1.3 Multipath Fading

Path loss and shadowing contribute to large scale fading, whereas the major reason behind small scale fading is multipath phenomena [72]. A *multipath* situation arises when propagation is mainly by way of scattering and several copies of the transmitted signal arrive sequentially from different directions at the receiver. Each version of the incoming radio wave will have different time delay due to difference in associated path length. The effect of these differential time delays is to introduce relative phase shifts between the component waves, often leading to destructive interference. The resultant signal obtained at receiver antenna consists of a superposition of different components and is usually a heavily attenuated version of the transmitted signal. It may happen that a receiver at a given location experiences a signal that is several tens of dB different from that at another location a short distance away where the phase relationships are different. This shows why multipath is referred to as small scale fading mechanism. Within a short range (< 1 m), the local mean value of the mobile radio signal is determined by the large scale fading effects and is nearly constant while the variation of the signal strength is caused by small scale fading which is superimposed on the large scale effects.

3.1.4 Time-variance of the Channel

Time variance of the channel refers to the change of the channel transfer function over time and is primarily attributed to the relative velocity between transmitter and receiver. In personal cellular communication, generally the base station (BS) is fixed and time variance is observed due to any movement of the mobile terminal.

Fading is basically a spatial phenomenon, but spatial variations are experienced as temporal variations by a receiver moving through the fading environment. When viewed in the frequency domain, relative velocity between transmitter and receiver manifests itself as *Doppler shift* [72].

3.1.5 Multipath Time-varying Channel

If we transmit an impulse over a multipath fading channel, the received signal appears as a train of pulses, as shown in Fig. 3.1. Let us assume that the impulses are transmitted at discrete time instants $\{t_i\}_{i=0}^2$, and the first copy of the transmitted signal is received at time $\{t_i'\}_{i=0}^2$. The first copy is the direct LOS component, if there exists any such path between transmitter and receiver. Otherwise, it refers to the scattered component having minimum path length. Further, we designate the excess delay of the jth replica, associated with the impulse transmitted at time t_i as τ_{ij}. In addition to the multipath phenomena, if the channel is affected by time variance too, then the nature of these multiple copies change with time.

Fig. 3.1 depicts that if we repeat pulse sounding experiment at a different time instant, we shall observe changes in the sizes of individual pulses, changes in the relative delays among pulses, and may be changes in the number of pulses. As the number of paths may be quite large and time varying, the fading caused by multipath is random in nature and can only be modelled statistically. The modelling and multipath channel impulse response is discussed in detail in Section 3.2.1.

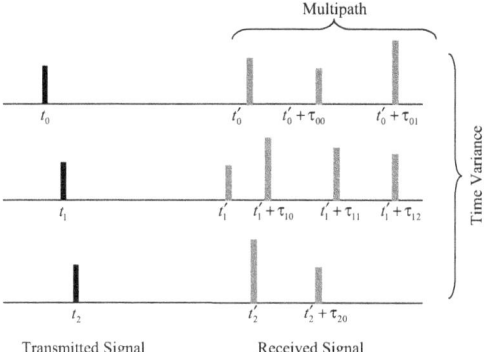

Figure 3.1: Impulse response of a multipath time-variant channel [2].

3.1.6 Flat and Frequency-Selective Fading

When a signal is transmitted through a fading channel if all the spectral components present in the received signal experience same attenuation and phase shift, the fading is said to be *frequency-nonselective* or equivalently *flat* fading. Flat fading is observed in case of narrowband signals, where the message signal bandwidth (W) is much smaller than the channel's coherence bandwidth (f_o). We define *coherence bandwidth* as the frequency range over which the mobile radio channel has a constant gain and linear phase response. The channel gain, however, may change with time [27].

To put it in simple terms, let us consider two impulses at transmitter end. The impulses are generated at same time but with different frequencies f_1 and f_2, where $\Delta f = |f_1 - f_2|$. Measurement of the channel response (at receiver) due to these impulses shows a degree of correlation. The correlation decreases as Δf increases and for $|f_1 - f_2| > \Delta f_{max}$ the correlation function value falls below a certain predetermined threshold value. Coherence bandwidth is simply equal to Δf_{max}.

On the other hand, if spectral components of the transmitted signal are affected by different amplitude gains and phase shifts, the fading is said to be *frequency selective*. Wideband signals with $W > f_o$ suffers from such fading.

One can have a deeper understanding of flat and frequency selective fading by viewing their difference in time domain. We define a parameter τ_m, which is related to the coherence bandwidth f_o as [73]

$$f_o \approx \frac{1}{\tau_m} \tag{3.1}$$

for the purpose. For a single transmitted impulse, τ_m denotes the multipath time spread, i.e. the time between the first and last received component. When the channel multipath delay spread is much smaller than the symbol duration $T_s (= 1/W)$

$$\tau_m < T_s \Rightarrow \frac{1}{f_o} < \frac{1}{W} \Rightarrow f_o > W \tag{3.2}$$

coherence bandwidth exceeds the signal bandwidth and the channel exhibits flat fading. In this case all the received multipath components of a symbol arrive within the symbol time duration and they are *non-resolvable*. For frequency selective channels

$$\tau_m > T_s \Rightarrow \frac{1}{f_o} > \frac{1}{W} \Rightarrow f_o < W \tag{3.3}$$

As $\tau_m > T_s$, the multiple copies that arrive at the receiver is not bounded by symbol time and affect adjacent symbols causing ISI. Equalizers are therefore needed for frequency selective channels.

Modelling of flat fading channels is relatively easy as non-resolvable multipath components give rise to amplitude variations only. Channel instantaneous gain is random and can be modeled with various probability distributions (Rayleigh, Rician etc.). On the contrary, modelling of frequency selective channels is quite involved. Each multipath signal has to be considered separately. Although some analytical models have been developed to characterize frequency selective fading, all of them are approximated heavily for tractability. One such model is two-ray Rayleigh fading model [27] where the channel response is made up of two delta functions which independently fade and have sufficient time delay between them to induce frequency selective fading. In comparison, the models based on wideband multipath measurements are flexible and accurate but are computationally intensive.

3.1.7 Slow and Fast Fading

If the attenuation and phase shift of any signal received through a fading channel remains constant over at least one symbol duration, the fading is said to be *slow*. This happens when the channel characteristics change less rapidly compared to the variations in the baseband transmitted signal. Mathematically, for slow fading the symbol duration (T_s) must be small enough compared to the coherence time (T_o). *Coherence time* can be defined as the width of a sliding time window over which fading channel exhibits constant gain and linear phase relationship.

Like coherence bandwidth, coherence time can also be explained with the help of two transmitted impulses. Here they should be generated at same frequency but at two different time t_1 and t_2. Let, $\Delta t = |t_1 - t_2|$ and Δt_{max} be the maximum value of Δt for which the impulse response correlation function value is maintained above some threshold.

The coherence time equals Δt_{max} for the given threshold value.

Fast fading occurs when symbol duration is larger than the coherence time, i.e. $T_s > T_o$, and the fading changes several times during any symbol transmission. Fast fading is observed for very low data rate signals (e.g. teletype or Morse code). However, most present-day terrestrial mobile radio channels can generally be characterized by slow fading channels. The discussion points to a lower limit ($R > 1/T_o$) on the signalling rate for slow fading approximation. It is interesting to note that a corresponding upper limit exist ($R < 1/\tau_m$) for flat fading approximation.

We begin the frequency domain characterization of slow and fast fading by stating the relation between coherence time and Doppler shift

$$T_o \approx \frac{1}{f_D} \tag{3.4}$$

where f_D is the Doppler shift caused by time variance of the channel. When the Doppler shift is much smaller than the signal bandwidth $W(= 1/T_s)$,

$$f_D < W \Rightarrow \frac{1}{T_o} < \frac{1}{T_s} \Rightarrow T_o > T_s \tag{3.5}$$

coherence time exceeds the symbol duration and the channel exhibits slow fading. In slow fading a particular fade level will affect many successive symbols, which leads to burst errors [73]. For fast fading channels

$$f_D > W \Rightarrow \frac{1}{T_o} > \frac{1}{T_s} \Rightarrow T_o < T_s \tag{3.6}$$

and the fading changes from symbol to symbol. This renders any analysis difficult when the communication receiver decisions are based on observations over two or more symbol durations (e.g. differentially coherent modulation or coded communications).

The Doppler shift is attributable to relative motion between transmitter and receiver. Hence, the velocity and baseband signaling rate determines whether a signal undergoes fast fading or slow fading.

3.1.8   Fading Mechanisms - A Summary

As this book is concerned with small scale fading effects on digital modulated signals, it would not be irrelevant to present a summary of different manifestations of small scale fading. Table 3.1 summarizes all of them along with the corresponding criteria (both in time and frequency domain).

There are two main reasons of signal degradation leading to small scale fading - multipath and time variance of the channel. Multipath causes delay spread which in turn results time dispersion or ISI whereas time variance causes Doppler shift or frequency dis-

Time Domain					Symbol	Description
Flat fading	$\tau_m < T_s$	Slow fading	$T_o > T_s$		T_s	symbol time
Frequency -selective fading	$\tau_m > T_s$	Fast fading	$T_o < T_s$		τ_m	maximum excess delay
					T_o	coherence time
Frequency Domain						
Flat fading	$f_o > W$	Slow fading	$f_D < W$		W	signal BW
Frequency -selective fading	$f_o < W$	Fast fading	$f_D > W$		f_o	coherence BW
					f_D	Doppler shift

Table 3.1: Small scale fading - time domain conditions, frequency domain conditions, and notations.

persion. The study can be done in either time or frequency domain. The time dispersion mechanism is characterized in the time domain as multipath delay spread, and in the frequency domain as channel coherence bandwidth. Similarly, the time-variant mechanism is characterized in the time domain as a channel coherence time, and in the Doppler-shift (frequency) domain as a channel fading rate or Doppler spread. Depending on the relation between the signal parameters (bandwidth, symbol duration) and the channel parameters (delay spread, Doppler shift) different type of fading manifestations are possible.

Finally we present a tree diagram showing all different kind of fading mechanisms in Fig. 3.2. The diagram includes the small scale fading types discussed above alongwith different large scale fading mechanisms. Path loss is characterized as attenuation with distance and shadowing is denoted with variation about the mean signal strength (as calculated from path loss).

3.2 Fading Models

3.2.1 Impulse Response of a Multipath Time-varying Channel

We assume a wireless environment where, apart from the direct LOS path, there are N distinct propagation paths between the transmitter and receiver. Also, let the transmitted signal $s(t)$ be a narrowband signal centered around carrier frequency f_c such that

$$s(t) = \Re\{\tilde{s}(t)\exp(j2\pi f_c t)\} \tag{3.7}$$

where $\tilde{s}(t)$ is the low-pass equivalent or complex envelope of $s(t)$.

Now in absence of noise, the received signal can be expressed as a summation of

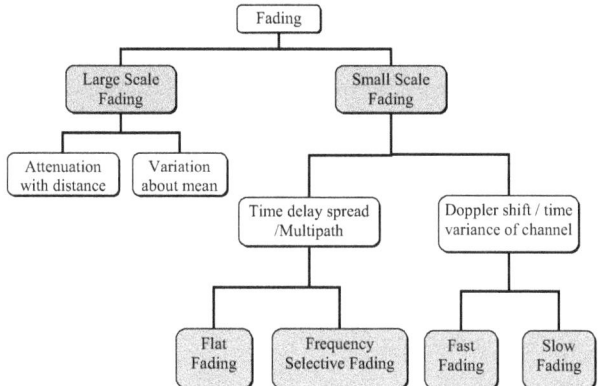

Figure 3.2: Different types of fading mechanisms.

multiple copies of $s(t)$ arriving through these paths as [74]

$$r(t) = \sum_{k=0}^{N} c_k s(t - T_k) \qquad (3.8)$$

where $k = 0$ denotes the LOS path, and other values of $k = 1, 2, \ldots, N$ accounts for the N equally spaced discrete multipath components. The attenuation factor and propagation delay associated with kth path are c_k and T_k respectively. If we assume a static multipath situation, these quantities can be regarded as constants. The magnitude of c_k depends on the cross sectional area of the kth reflecting surface or the length of the kth diffracting edge. On the other hand, propagation delay $T_k = l_k/v_c$ depends on path length l_k, where v_c is velocity of electromagnetic waves. Further, T_k can be expressed as $T_k = T_0 + \tau_k$, T_0 being the delay associated with the LOS path whereas τ_k is the excess delay. By excess delay we mean the relative delay of the kth multipath component as compared to the LOS component. Quite naturally, $\tau_0 = 0$. Thus the received signal $r(t)$ given in (3.8) may also be expressed as

$$r(t) = \sum_{k=0}^{N} c_k s(t - T_0 - \tau_k) \qquad (3.9)$$

As c_k and T_k are both real, from (3.7) and (3.9), we have

$$r(t) = \Re\left\{\sum_{k=0}^{N} c_k \tilde{s}(t - T_0 - \tau_k) \exp[j2\pi f_c(t - T_0 - \tau_k)]\right\} \tag{3.10}$$

The received signal $r(t)$ is also a bandpass signal and can be expressed with its lowpass equivalent (see Section 2.2.1 for details), $r(t) = \Re\{\tilde{r}(t)\exp(j2\pi f_c t)\}$. As we are only interested in the effect of excess delay introduced by multipath components, the lowpass equivalent can be expressed in the following way to absorb the propagation delay term

$$r(t) = \Re\{\tilde{r}(t - T_0)\exp[j2\pi f_c(t - T_0)]\} \tag{3.11}$$

Comparing (3.10) and (3.11), we get

$$\tilde{r}(t) = \sum_{k=0}^{N} c_k \tilde{s}(t - \tau_k)\exp(-j2\pi f_c \tau_k) = \sum_{k=0}^{N} c_k \tilde{s}(t - \tau_k)\exp(-j\phi_k) \tag{3.12}$$

where $\phi_k = 2\pi f_c \tau_k$. This is an important result because it tells us that even though a bandpass signal $r(t)$ experiences the fading effect, these effects can be analyzed at the baseband level.

Impulse Response

If $\tilde{S}(f)$ be the equivalent lowpass spectrum of the complex envelope of the transmitted signal $\tilde{s}(t)$, i.e.

$$\tilde{s}(t) = \int_{-\infty}^{\infty} \tilde{S}(f)\exp(j2\pi ft)df \tag{3.13}$$

then we may write from (3.7) and (3.13)

$$s(t) = \Re\left\{\exp(j2\pi f_c t)\int_{-\infty}^{\infty} \tilde{S}(f)\exp(j2\pi ft)df\right\} \tag{3.14}$$

Similarly the complex envelope of the received signal can be expressed as

$$\tilde{r}(t) = \int_{-\infty}^{\infty} \tilde{R}(f)\exp(j2\pi ft)df = \int_{-\infty}^{\infty} \tilde{S}(f)\tilde{H}(f)\exp(j2\pi ft)df \tag{3.15}$$

where $\tilde{R}(f) = \tilde{S}(f)\tilde{H}(f)$ [4, (A.16)] and $\tilde{R}(f)$, $\tilde{H}(f)$ are lowpass spectrum of the received signal and lowpass equivalent channel transfer function respectively.

On the other hand, from (3.12) and (3.13), another expression of $\tilde{r}(t)$ can be obtained

$$
\begin{aligned}
\tilde{r}(t) &= \sum_{k=0}^{N} c_k \left\{ \int_{-\infty}^{\infty} \tilde{S}(f) \exp[j2\pi f(t - \tau_k)] df \right\} \exp(-j2\pi f_c \tau_k) \\
&= \int_{-\infty}^{\infty} \tilde{S}(f) \exp(j2\pi ft) \left\{ \sum_{k=0}^{N} c_k \exp[-j2\pi(f + f_c)\tau_k] \right\} df
\end{aligned}
\tag{3.16}
$$

Now, comparing (3.15) and (3.16), we have

$$
\tilde{H}(f) = \sum_{k=0}^{N} c_k \exp[-j2\pi(f + f_c)\tau_k]
\tag{3.17}
$$

The inverse Fourier transform of $\tilde{H}(f)$ yields the lowpass equivalent channel impulse response

$$
\tilde{h}(t) = \sum_{k=0}^{N} c_k \exp(-j2\pi f_c \tau_k)\delta(t - \tau_k)
\tag{3.18}
$$

where $\delta(t)$ is the Dirac-delta function and $\tilde{r}(t)$ is the discrete convolution of $\tilde{s}(t)$ and $\tilde{h}(t)$, i.e. $\tilde{r}(t) = \tilde{s}(t) * \tilde{h}(t)$. Note that the time shifting property of Fourier transform has been used to derive (3.18).

In dynamic multipath situations, attenuation factor as well as propagation delay become time variant due to the time varying nature of the channel. In such an environment, (3.12) can be written as

$$
\begin{aligned}
\tilde{r}(t) &= \sum_{k=0}^{N} c_k \tilde{s}[t - \tau_k(t)] \exp[-j2\pi f_c \tau_k(t)] \\
&= \sum_{k=0}^{N} c_k \tilde{s}[t - \tau_k(t)] \exp[-j\phi_k(t)]
\end{aligned}
\tag{3.19}
$$

where $\phi_k(t) = 2\pi f_c \tau_k(t)$. Since the carrier frequency f_c is very large, very small changes in the path delays $\tau_k(t)$ will cause large changes in the phase $\phi_k(t)$. For example, a 900 MHz sinusoid has a wavelength of about 30 cm. Since, radio waves propagate at about 30 cm per nanosecond (ns), a path delay change of just 1 ns corresponds to one full wavelength (or 2π radians phase shift) in the 900 MHz sinusoid. Some authors [4] even consider the number of multipaths to be time varying, i.e. they use $N(t)$ instead of N. However we have assumed the number of multipaths to be static to avoid further complexity in analysis.

For time-varying channel the lowpass channel transfer function may be obtained by introducing additional time dependence in (3.17)

$$\tilde{H}(f,t) = \sum_{k=0}^{N} c_k(t) \exp[-j2\pi(f + f_c)\tau_k(t)] \tag{3.20}$$

The lowpass channel impulse response is the inverse Fourier transform of $\tilde{H}(f,t)$. However, as the channel spectrum is itself a function of time, we require two variables to denote time. The channel impulse response $\tilde{h}(\tau,t)$, i.e., the channel output at time t in response to an impulse applied to the channel at $t - \tau$, is thus

$$
\begin{aligned}
\tilde{h}(\tau,t) &= \sum_{k=0}^{N} c_k(t) \exp[-j2\pi f_c\tau_k(t)]\delta[\tau - \tau_k(t)] \\
&= \sum_{k=0}^{N} c_k(t) \exp[-j\phi_k(t)]\delta[\tau - \tau_k(t)]
\end{aligned}
\tag{3.21}
$$

Equation (3.21) conforms to the linear time-variant (LTV) channel model as the impulse response is characterized as a function of two variables, one describing the instant of observing the channel output t, and the other describing the propagation delay τ [59].

Flat Fading Assumption

If the differential path delays $\tau_i - \tau_j$ are small compared to the duration of a modulated symbol, then all the $\tau_k(t)$ terms in (3.21) may be assumed approximately equal to each other, i.e. $\tau_k = \hat{\tau}; \forall k$. In this case, the channel impulse response has the form [75]

$$
\begin{aligned}
\tilde{h}(\tau,t) &= \sum_{k=0}^{N} c_k(t) \exp[-j\phi_k(t)]\delta[\tau - \hat{\tau}(t)] \\
&= \alpha(t) \exp[-j\phi(t)]\delta[\tau - \hat{\tau}(t)]
\end{aligned}
\tag{3.22}
$$

The corresponding channel transfer function is obtained by taking the Fourier transform of $\tilde{h}(\tau,t)$, giving

$$\tilde{H}(f,t) = \alpha(t) \exp[-j\phi(t)] \exp(-j2\pi ft) \tag{3.23}$$

Since the channel amplitude response is $|\tilde{H}(f,t)| = \alpha(t)$, all frequency components in the received signal are subjected to the same complex gain $\alpha(t)$, and the channel is said to exhibit flat fading. The fading takes the form of random attenuation as the transmitted signal is multiplied by a random value. For obvious reasons, it is sometimes called *multiplicative fading*.

Narrowband Fading Model

For narrowband signals the frequency content is concentrated in the vicinity of the carrier frequency f_c. Equivalently, in the baseband, the frequency content of $\tilde{s}(t)$ is concentrated around $f = 0$. This implies that within the bandwidth occupied by $\tilde{S}(f)$, the time variant channel transfer function $\tilde{H}(f,t)$ is a complex valued constant in the frequency variable [2]. Since $\tilde{S}(f)$ has its frequency content concentrated in the vicinity of $f = 0$, $\tilde{H}(f,t) = \tilde{H}(0,t)$. Now, from (3.15), we have

$$
\begin{aligned}
\tilde{r}(t) &= \int_{-\infty}^{\infty} \tilde{S}(f)\tilde{H}(f,t)\exp(j2\pi ft)df \\
&= \tilde{H}(0,t)\int_{-\infty}^{\infty} \tilde{S}(f)\exp(j2\pi ft)df
\end{aligned}
\tag{3.24}
$$

and further using the expression of $\tilde{s}(t)$ given in (3.13)

$$
\tilde{r}(t) = \tilde{H}(0,t)\tilde{s}(t)
\tag{3.25}
$$

Finally putting $f = 0$ in (3.23) and substituting the same in (3.25) we get

$$
\tilde{r}(t) = \alpha(t)\exp[-j\phi(t)]\tilde{s}(t) + \tilde{n}(t)
\tag{3.26}
$$

where $\alpha(t)$ and $\phi(t)$ are resultant amplitude and phase terms, while $\tilde{n}(t)$ accounts for the additional term.

Slow Fading Assumption

We define the fading to be slow when the attenuation and phase shift of the received signal remain constant over one symbol duration. In this case we may ignore the time variations of the random processes $c_k(t)$ and $\phi_k(t)$ and replace them with random variables c_k and ϕ_k, where c_k is an independent and identically distributed (i.i.d.) RV and ϕ_k is uniformly distributed over the interval $[-\pi, \pi]$.

Thus under slow and flat fading, and when the transmitted signal is narrowband, the baseband received signal can be expressed as

$$
\tilde{r}(t) = \alpha\exp(-j\phi)\tilde{s}(t) + \tilde{n}(t)
\tag{3.27}
$$

3.2.2 Some Useful Transformations

When fading affects narrowband systems, the received carrier amplitude is modulated by the fading amplitude α, where α is a RV dependent on the nature of the radio propagation environment. In addition to fading the signal is also perturbed by AWGN, which is typically assumed to be statistically independent of the fading amplitude α. As the parameter α represents instantaneous signal envelope, α^2 denotes instantaneous attenuation in received signal power. Thus, in a fading environment, we define the instantaneous SNR per symbol by $\gamma = \alpha^2 \eta; \eta = E_s/N_0$ and the average SNR per symbol by $\bar{\gamma} = E\{\gamma\} = \Omega \eta$, where $\Omega = E\{\alpha^2\} = \bar{\alpha}^2$ is the average fading power. Quite evidently in the absence of fading mechanism, instantaneous SNR per symbol is simply $\gamma = \eta$, while for binary signalling $\gamma_b = \eta_b$ and $\bar{\gamma}_b = \Omega \eta_b$ respectively.

From the relations $\gamma = \alpha^2 \eta$ and $\bar{\gamma} = \Omega \eta$ we can write $\gamma = \alpha^2 \bar{\gamma}/\Omega$ or $\alpha = \sqrt{\gamma \Omega/\bar{\gamma}}$. Again starting from $\gamma = \alpha^2 \eta$ and differentiating the same, we get $d\gamma = 2\alpha \eta d\alpha = 2\eta \sqrt{\gamma \Omega/\bar{\gamma}} d\alpha = 2\sqrt{\gamma \bar{\gamma}/\Omega} d\alpha$. Now from the relation

$$f_\alpha(\alpha)d\alpha = f_\gamma(\gamma)d\gamma \tag{3.28}$$

the PDF $f_\gamma(\gamma)$ may be obtained by the method of change of RV as [73]

$$f_\gamma(\gamma) = \frac{f_\alpha(\sqrt{\gamma \Omega/\bar{\gamma}})}{2\sqrt{\gamma \bar{\gamma}/\Omega}} \tag{3.29}$$

It may be noted that (3.28) follows from the fact that there exists an one-to-one relation between the RVs. i.e. $F_\gamma(\gamma_0) = \Pr[\gamma < \gamma_0] = \Pr[\alpha < \alpha_0; \alpha_0 = g^{-1}(\gamma_0)] = F_\alpha(\alpha_0)$, where $\gamma = g(\alpha)$ and $F_z(\cdot)$ denotes CDF of z.

Another important statistical characteristic of fading channels is the moment generating function (MGF) defined as [73]

$$M_\gamma(s) = \int_0^\infty f_\gamma(\gamma) \exp(-s\gamma) d\gamma \tag{3.30}$$

Taking the first derivative of (3.30) with respect to s and evaluating the result at $s = 0$, we get

$$\bar{\gamma} = -\frac{dM_\gamma(s)}{ds}\bigg|_{s=0} \tag{3.31}$$

as $\bar{\gamma} = \int_0^\infty \gamma f_\gamma(\gamma) d\gamma$ by definition. This shows how average SNR can be easily calculated with the help of MGF. Next, we show that how MGF serves as a useful tool in evaluating the error probability expressions of different modulation schemes in fading environment.

For a digital receiver, which makes its decision based on an undistorted symbol waveform corrupted by stationary AWGN only, the SEP depends only on the instantaneous SNR η associated with each symbol. The SEP in a fading channel, on the other hand, becomes a conditional error probability (CEP) $P_e(\gamma)$ conditioned by γ, and the corresponding average SEP can be found by averaging the CEP over γ as

$$P_e = \int_0^\infty P_e(\gamma) f_\gamma(\gamma) d\gamma \tag{3.32}$$

where $f_\gamma(\gamma)$ is the PDF of γ for a specified fading environment, and $P_e(\gamma)$ refers to the SEPs described in Chapter 1.3 as a function of η, with γ replacing η in each equation. This method is popularly referred to as the standard PDF approach in the literature. If the CEP can be expressed in an exponential form

$$P_e(\gamma) = C_1 \exp(-C_2\gamma) \tag{3.33}$$

the average SEP P_e may be found easily through MGF method as

$$P_e = C_1 M_\gamma(C_2) \tag{3.34}$$

avoiding the need of performing an integration with infinite limits. The corresponding method is known as MGF method and heavily practiced in current publications [73].

Finally it may be noted that the conventional definition of MGF is $M_\gamma(s) = \int_0^\infty f_\gamma(\gamma) \exp(s\gamma) d\gamma$ i.e. the Laplace transform of the PDF $f_\gamma(\gamma)$ with the argument reversed in sign, $L\{f_\gamma(\gamma)\} = M_\gamma(-s)$. However, due to the presence of negative exponentials in various conditional error probability expressions, the definition in (3.30) is generally preferred.

3.2.3 Rayleigh Fading

Under the assumptions of flat and slow fading, the complex channel gain for a fading channel is given by $\sum_{k=0}^N c_k \exp(-j\phi_k)$. The worst case scenario will arise when the direct LOS component is absent and only scattered replicas are available, the corresponding the channel gain being $\sum_{k=1}^N c_k \exp(-j\phi_k) = \alpha \exp(-j\phi)$. The received signal envelope in this case (in absence of noise) will be $\tilde{r}(t) = \sum_{k=1}^N c_k \exp(-j\phi_k) \tilde{s}(t) = \alpha \exp(-j\phi) \tilde{s}(t)$.

Let $X = \sum_{k=1}^N c_k \exp(-j\phi_k)$ be a sum of complex random numbers denoting random amplitudes and phases. From central limit theorem, we can say that if N is large enough then this sum is well approximated by complex Gaussian PDF [3]. In other words, if X is resolved into real and imaginary components $X = X_1 + jX_2$, then both X_1 and X_2 become Gaussian RVs with mean 0 and equal variance $\Omega/2$, i.e. $\{X_1, X_2\} \sim \mathcal{N}(0, \Omega/2)$ as

$$E\{X\} = E\left\{\sum_{k=1}^N c_k \exp(-j\phi_k)\right\} = \sum_{k=1}^N E\{c_k\} E\{\exp(-j\phi_k)\} = 0 \tag{3.35}$$

when $\phi_k \sim U[-\pi, \pi]$ and

$$E\{X^2\} = \sum_{k=1}^{N} E\{c_k^2\} = \bar{\alpha^2} = \Omega \tag{3.36}$$

If $\{X_1, X_2\}$ follows Gaussian distribution, then it can be shown that [18] the corresponding fading amplitude $\alpha = \sqrt{X_1^2 + X_2^2}$ would follow a Rayleigh distribution

$$f_\alpha(\alpha) = \frac{2\alpha}{\Omega} \exp\left(-\frac{\alpha^2}{\Omega}\right) \qquad ; \quad \alpha \geq 0 \tag{3.37}$$

where $E\{\alpha^2\} = \Omega$. Integrating $f_\alpha(\alpha)$ given in (3.37) with respect to α, the corresponding CDF may be found as

$$F_\alpha(\alpha) = 1 - \exp\left(-\frac{\alpha^2}{\Omega}\right) \qquad ; \quad \alpha \geq 0 \tag{3.38}$$

From (3.29) and (3.37) the PDF of average SNR (γ) under Rayleigh fading is given by

$$f_\gamma(\gamma) = \frac{1}{\bar{\gamma}} \exp\left(-\frac{\gamma}{\bar{\gamma}}\right) \qquad ; \quad \gamma \geq 0 \tag{3.39}$$

which is exponentially distributed. Integrating $f_\gamma(\gamma)$ given in (3.39) with respect to γ, we get the CDF of γ

$$F_\gamma(\gamma) = 1 - \exp\left(-\frac{\gamma}{\bar{\gamma}}\right) \qquad ; \quad \gamma \geq 0 \tag{3.40}$$

Finally the MGF corresponding to this fading model may be calculated as

$$M_\gamma(s) = \frac{1}{1 + s\bar{\gamma}} \tag{3.41}$$

3.2.4   Rician Fading

When there are a large number of scatterers and these scatterers are randomly moving, the impulse response $\tilde{h}(\tau, t)$ can be modeled as zero mean complex valued Gaussian process and the envelope $|\tilde{h}(\tau, t)|$ at any instant t is Rayleigh distributed. Rayleigh distribution can be characterized with a single parameter $E\{\alpha^2\} = \Omega$. But when there are fixed scatterers and reflectors, an LOS path is present and $\tilde{h}(\tau, t)$ can no longer be modeled as zero mean. In this case $|\tilde{h}(\tau, t)|$ has a Rician distribution with two parameters s and Ω, where s^2 represents the power in the non-fading signal component (specular component). Quite naturally, the fading in this case is less severe due to the presence of specular component.

66

Substituting $\phi_0(t) = 0$ (as excess delay for the LOS component $\tau_0(t) = 0$) in (3.19), the received signal envelope is

$$\tilde{r}(t) = c_0(t)\tilde{s}(t) + \sum_{k=1}^{N} c_k(t)\tilde{s}[t - \tau_k(t)]\exp[-j\phi_k(t)] \tag{3.42}$$

For unmodulated carrier or narrowband signal under flat fading assumption

$$\tilde{r}(t) = c_0(t) + \sum_{k=1}^{N} c_k(t)\exp[-j\phi_k(t)] \tag{3.43}$$

and under slow fading assumption,

$$\tilde{r}(t) = c_0 + \sum_{k=1}^{N} c_k \exp(-j\phi) \tag{3.44}$$

The envelope has the same form for a Rayleigh fading channel except the specular component given by $s = c_0$. Application of central limit theorem (when N is large) enables us to characterize $\tilde{r}(t)$ as a complex Gaussian process $\tilde{r}(t) = X_1 + jX_2$ where $X_1 \sim \mathcal{N}(s, \sigma^2)$ and $X_2 \sim \mathcal{N}(0, \sigma^2)$. The real part has a non-zero mean which accounts for the specular component. The average power of the non-LOS multipath components $\sum_{k, k \neq 0} E\{c_k^2\} = 2\sigma^2$ and $s^2 = c_0$ is the power in the LOS component. The total average power is $E\{\alpha^2\} = s^2 + 2\sigma^2 = \Omega$.

Accordingly, the fading amplitude $\alpha = \sqrt{X_1^2 + X_2^2}$ follows a Rician distribution [18] and is given by

$$f_\alpha(\alpha) = \frac{\alpha}{\sigma^2}\exp\left(-\frac{\alpha^2 + s^2}{2\sigma^2}\right)I_0\left(\frac{\alpha s}{\sigma^2}\right) \qquad ; \quad \alpha \geq 0 \tag{3.45}$$

The Rician channel has an important parameter called the K factor, defined by $K = s^2/(2\sigma^2)$, i.e. the ratio of the power of the LOS component to the total power of all other scattered components. Rician distribution is a more general distribution and it includes Rayleigh distribution as a special case when $K = 0$, i.e. when no LOS component is present. On the other hand, as $K \to \infty$, only the dominant component matters and there is no fading. As a result, the wireless channel becomes an AWGN channel. The fading parameter K is therefore a measure of severity of the fading; a small K implies severe fading, and a large K value implies milder fading. Fig. 3.3 shows the Rician distribution with $\sigma = 1$ for various K values.

The Rice distribution is also known as the Nakagami-n distribution where n is the Nakagami-n fading parameter, which ranges from 0 to ∞. This parameter is related to the Rician K factor by $K = n^2$. From the relations $s^2 + 2\sigma^2 = \Omega$ and $K = s^2/(2\sigma^2)$, we can write $s = \sqrt{K\Omega/(1+K)}$ and $\sigma^2 = \Omega/[2(1+K)]$. Substituting these values in (3.45)

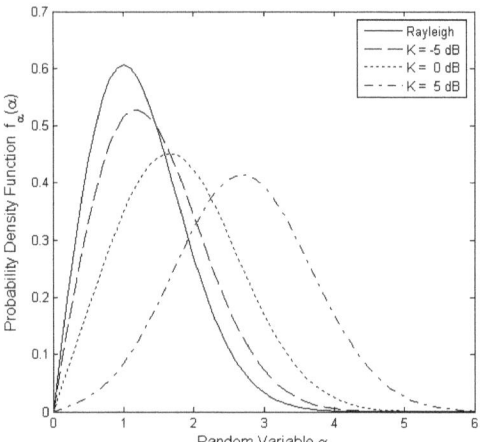

Figure 3.3: Rician fading distribution for $\sigma = 1$.

the PDF becomes

$$f_\alpha(\alpha) = \frac{2\alpha(1+K)}{\Omega} \exp\left[-K - \frac{\alpha^2(1+K)}{\Omega}\right] I_0 \left[2\alpha\sqrt{\frac{K(1+K)}{\Omega}}\right] \tag{3.46}$$

or in terms of n,

$$f_\alpha(\alpha) = \frac{2\alpha(1+n^2)}{\Omega} \exp\left[-n^2 - \frac{\alpha^2(1+n^2)}{\Omega}\right] I_0 \left[2\alpha n\sqrt{\frac{1+n^2}{\Omega}}\right] \tag{3.47}$$

for $\alpha \geq 0$. Integrating $f_\alpha(\alpha)$ with respect to α and using the definition of Marcum's Q function [36], we obtain the CDF of α

$$F_\alpha(\alpha) = 1 - Q_1\left(\sqrt{2K}, \alpha\sqrt{\frac{2(1+K)}{\Omega}}\right) \qquad ; \quad \alpha \geq 0 \tag{3.48}$$

Using (3.29) the PDF of γ under Rician fading (for $\gamma \geq 0$)

$$f_\gamma(\gamma) = \frac{1+K}{\bar{\gamma}} \exp\left[-K - \frac{\gamma(1+K)}{\bar{\gamma}}\right] I_0 \left[2\sqrt{\frac{\gamma K(1+K)}{\bar{\gamma}}}\right] \tag{3.49}$$

which is a noncentral chi-square distribution with $n = 2$ degrees of freedom. The corresponding CDF is

$$F_\gamma(\gamma) = 1 - Q_1\left(\sqrt{2K}, \sqrt{\frac{2(1+K)\gamma}{\bar{\gamma}}}\right) \qquad ; \quad \gamma \geq 0 \tag{3.50}$$

and the MGF corresponding to this fading model is given by

$$M_\gamma(s) = \frac{1+K}{1+K+s\bar{\gamma}} \exp\left(-\frac{Ks\bar{\gamma}}{1+K+s\bar{\gamma}}\right) \tag{3.51}$$

3.2.5 Nakagami Fading

Some experimental data does not fit well into either Rayleigh or Rician distribution. For this reason a more general distribution describing the envelope of fading signal was developed. It is called Nakagami (or more specifically Nakagami-m) distribution and is defined by

$$f_\alpha(\alpha) = \frac{2}{\Gamma(m)}\left(\frac{m}{\Omega}\right)^m \alpha^{2m-1} \exp\left(-\frac{m\alpha^2}{\Omega}\right) \qquad ; \quad \alpha \geq 0 \tag{3.52}$$

Nakagami distribution often gives the best fit for land mobile and indoor mobile multipath propagation, as well as for scintillating ionosphere radio links [73]. Integrating (3.52) with respect to α and using the definition of complementary incomplete gamma function $\Gamma(\cdot,\cdot)$ [8, (8.350.2)] the corresponding CDF may be found as

$$F_\alpha(\alpha) = 1 - \frac{\Gamma(m, m\alpha^2/\Omega)}{\Gamma(m)} \qquad ; \quad \alpha \geq 0 \tag{3.53}$$

Nakagami distribution is a two parameter distribution involving the second moment $\Omega = E\{\alpha^2\}$ and fading figure $m = \Omega^2/E\{(\alpha^2 - \Omega)^2\}$ where $1/2 < m < \infty$ and $\Omega/2$ is the average power of the signal [76]. As a consequence, the distribution provides more flexibility and accuracy in matching the observed signal statistics. Fig. 3.4 shows the Nakagami distribution with $\Omega = 1$ for various m values.

For $m = 1$, the distribution reduces to Rayleigh fading PDF described in (3.37). Thus the Nakagami distribution can be used to model fading conditions that are either more or less severe than the Rayleigh distribution, and it includes the Rayleigh distribution as a special case ($m = 1$). For $m = 1/2$ it represents the one-sided Gaussian PDF with $\sigma = \sqrt{\Omega}$, for the range $1/2 \leq m < 1$ we have deeper fading than the Rayleigh, for $m > 1$ we have shallower fading than the Rayleigh (like Rician fading), and as $m \to \infty$ we approach non fading AWGN channel [76, 77].

The PDF for γ can be evaluated in a way that is similar to the derivation in case of

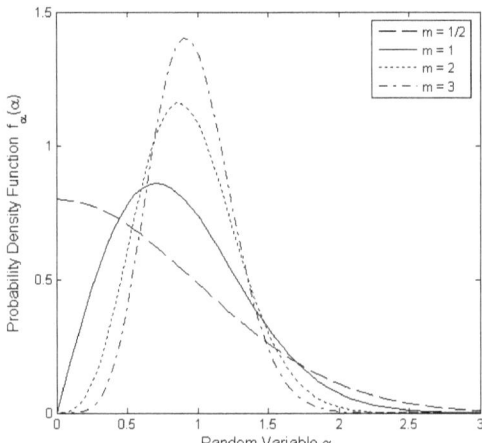

Figure 3.4: Nakagami fading distribution for $\Omega = 1$.

Rayleigh or Rician fading

$$f_\gamma(\gamma) = \frac{1}{\Gamma(m)} \left(\frac{m}{\bar{\gamma}}\right)^m \gamma^{m-1} \exp\left(-\frac{m\gamma}{\bar{\gamma}}\right) \quad ; \quad \gamma \geq 0 \tag{3.54}$$

which is a special case of the standard gamma distribution $f_x(x|a,b) = 1/[b^a\Gamma(a)]x^{a-1}\exp(-x/b)$ with parameters $a = m, b = \bar{\gamma}/m$. Further, the CDF of γ, $F_\gamma(\gamma)$ is given by

$$F_\gamma(\gamma) = 1 - \frac{\Gamma(m, m\gamma/\bar{\gamma})}{\Gamma(m)} \quad ; \quad \gamma \geq 0 \tag{3.55}$$

whereas the MGF is

$$M_\gamma(s) = \left(\frac{m}{m + s\bar{\gamma}}\right)^m \tag{3.56}$$

3.2.6 Hoyt Fading

Hoyt or Nakagami-q distribution is generally used to characterize the fading environments that are more severe than Rayleigh fading. The corresponding PDF of the fading envelope is given by

$$f_\alpha(\alpha) = \frac{(1+q^2)\,\alpha}{q\Omega} \exp\left[-\frac{(1+q^2)^2\alpha^2}{4q^2\Omega}\right] I_0 \left[\frac{(1-q^4)\alpha^2}{4q^2\Omega}\right] \tag{3.57}$$

where $\alpha \geq 0$ and q is the fading parameter, which ranges from 0 to 1. If we define $p = 4q^2/\left(1+q^2\right)^2 ; 0 \leq p \leq 1$, (3.57) may be expressed in a more compact form as

$$f_\alpha(\alpha) = \frac{2\alpha}{\sqrt{p}\Omega} \exp\left(-\frac{\alpha^2}{p\Omega}\right) I_0\left(\frac{\alpha^2\sqrt{1-p}}{p\Omega}\right) \qquad ; \quad \alpha \geq 0 \qquad\qquad (3.58)$$

Fig. 3.5 shows the Hoyt distribution with $\Omega = 1$ for various q values. For $q = 1$, the distribution reduces to Rayleigh fading PDF described in (3.37) and for $q = 0$ it represents the one-sided Gaussian PDF.

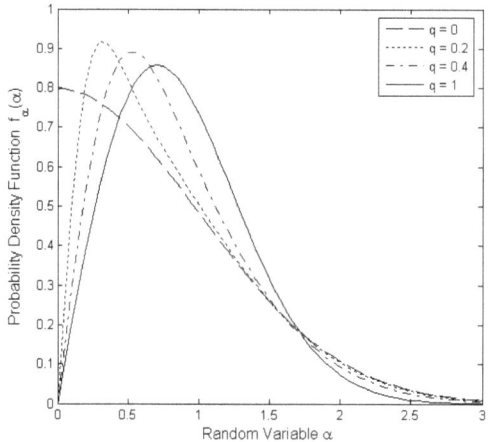

Figure 3.5: Hoyt fading distribution for $\Omega = 1$.

Using (3.29), the PDF of instantaneous SNR under Hoyt fading may be obtained from (3.57) as

$$f_\gamma(\gamma) = \frac{1+q^2}{2q\bar{\gamma}} \exp\left[-\frac{(1+q^2)^2\gamma}{4q^2\bar{\gamma}}\right] I_0\left[\frac{(1-q^4)\gamma}{4q^2\bar{\gamma}}\right] \quad ; \quad \gamma \geq 0 \qquad\qquad (3.59)$$

or equivalently

$$f_\gamma(\gamma) = \frac{1}{\sqrt{p}\bar{\gamma}} \exp\left(-\frac{\gamma}{p\bar{\gamma}}\right) I_0\left(\frac{\gamma\sqrt{1-p}}{p\bar{\gamma}}\right) \qquad ; \quad \gamma \geq 0 \qquad\qquad (3.60)$$

Expressions for the corresponding CDFs, i.e. $F_\alpha(\alpha)$ and $F_\gamma(\gamma)$, are not available in the open technical literature. The CDFs are derived in Chapter 6.3.1 and the associated derivations are thoroughly described there. The expression of MGF is, however, available in closed form and is given by

$$M_\gamma(s) = \left[1 + 2s\bar{\gamma} + \left(\frac{2sq\bar{\gamma}}{1+q^2} \right)^2 \right]^{-1/2} \qquad (3.61)$$

or equivalently

$$M_\gamma(s) = \frac{1}{\sqrt{1 + 2s\bar{\gamma} + ps^2\bar{\gamma}^2}} \qquad (3.62)$$

3.2.7 Applicability of Different Fading Models

In compliance with the random nature of fading in cellular mobile systems, statistical modelling approach is most suitable to characterize the fading effects. Depending on the particular propagation environment and the underlying communication scenario, several such models have been devised. Table 3.2 lists the four major statistical small-scale fading models (discussed in Section 3.2.1) alongwith their application environments.

Channel Type	Environment
Rayleigh	Mobile systems with no LOS path, propagation of reflected and refracted paths through troposphere and ionosphere, maritime ship-to-ship communication links
Rice (Nakagami-n)	LOS paths of microcellular urban and suburban land mobile, picocellular indoor, and factory environments, dominant LOS path of satellite radio links
Nakagami-m	Land mobile and indoor mobile multipath propagation, scintillating ionosphere radio links
Hoyt (Nakagami-q)	Satellite links subject to strong ionospheric scintillation

Table 3.2: Statistical models for various wireless environments.

The Rayleigh distribution is used to model the propagation environment where the mobile antenna receives a large number of reflected and scattered waves. The Nakagami-n distribution, known as the Rice distribution, is often used to model environments similar to Rayleigh fading channels, except that the set of reflected and scattered waves are dominated by one strong component. The Nakagami-m distribution can be used to model fading channel conditions that are more severe than the Rayleigh distribution. It often gives the best fit to land mobile as well as indoor mobile multipath propagation. The Nakagami-q distribution, also known as Hoyt distribution, is typically observed on satellite links subjected to strong ionospheric scintillation.

Statistical models are most accurate in environments with fairly regular geometries and uniform dielectric properties. Indoor environments tend to be less regular than outdoor environments, since the geometric and dielectric characteristics change dramatically depending on whether the indoor environment is an open factory, cubicled office, or metal machine shop. For these environments computer-aided modelling tools are available to predict signal propagation characteristics.

3.3 Diversity Techniques

3.3.1 Fading Mitigation

Many advanced signal transmission and processing methods have been developed for wireless systems to contravene the effect of fading. For example, wideband signalling techniques such as spread spectrum are often used as a countermeasure to frequency selective fading. Similarly, by dividing a high-rate signal to many parallel low-rate signals, orthogonal frequency division multiplexing (OFDM) mitigates the effect of channel dispersion. Equalization is another technique that is robust against channel dispersion. Frequency selective channel introduces different attenuation and phase shift to different frequency components in transmitted signal. An equalizer makes a frequency selective channel to appear as a flat fading channel by doing the opposite. On the other hand, fast fading occurs in low data rate transmission. If symbol duration is reduced by signal redundancy to a value lesser than the coherence time, the channel appears as a slow fading channel. Robust non-coherent and differentially coherent modulation schemes that avoid phase tracking and the associated delay involved or Doppler diversity (described later) are also effective against fast fading. Channel coding improves link performance by adding redundant data bits in the transmitted message so that if an instantaneous fade occurs in the channel, the data may still be recovered. For a given SNR, with coding present, the error floor out of the demodulator will not be lowered, but a lower error rate out of the decoder can be achieved. However, to combat slow flat fading which is the most common type encountered in wireless communications, the best known and most widely used techniques are the multi-channel combining methods referred to collectively as diversity.

3.3.2 Basic Diversity Concept

Diversity means the state of being composed of different parts, elements, or individuals. In the context of mobile communications, it implies the use of more than one copies of the signal to reduce fading.

Diversity improves transmission performance by making use of more than one independently faded version of the transmitted signal. If several replicas of the signal, carrying the same information, are received over multiple channels that exhibit independent fading with comparable strengths, the chances that all the independently faded signal components experience deep fading simultaneously are very less. Let p denote the probability of losing link connectivity or outage on each diversity branch. We say a link is in outage when the instantaneous SNR of the link is below a critical threshold and it cannot support the minimum QoS criteria. If fading occurs independently in all L diversity branches then the probability of outage on all channels simultaneously is p^L. Thus, a 10% chance of losing the signal for one channel is reduced to $0.1^3 = 0.001 = 0.1\%$ with three independently fading channels [78].

To achieve a high degree of improvement from a diversity system, the fading in individual branches should have low cross-correlation and the mean power available from each branch should be almost equal. If the cross-correlation is too high, then fades in each branch will occur simultaneously. On the other hand, if the branches have low correlation but have very different mean powers, then the signal in a weaker branch may not be useful even though it has less fades than the other branches.

3.3.3 Advantages of Diversity

The primary benefit of a diversity system over a non-diversity receiver can be summed up with diversity gain, defined as the decrease in required SNR with respect to the no-diversity case for a given QoS metric (such as BER) [79]. When diversity is realized with multiple antennas at the receiver, the receiver benefits from the array gain too. Array gain, does not rely on statistical independence between the channels, and instead, achieves its performance enhancement by coherently combining the energy received by each of the antennas. Even if the channels are completely correlated, as might happen in a LOS system, the received SNR increases linearly with the number of receive antennas, owing to the array gain [80].

Diversity also improves the data rate and hence the capacity of a wireless communication system. With receive diversity the average received SNR is increased at best linearly, owing to the array gain. The corresponding increase in data rate is given by the Shannon Hartley law, invoked in Section 2.4

$$C = B \log_2(1 + \gamma) \tag{3.63}$$

Since antenna diversity increases the SNR linearly, the capacity is enhanced logarithmically with respect to the number of antennas. In other words, the data rate benefit rapidly diminishes as antennas are added. However, it can be noted that when the SNR is low, the capacity increase is close to linear with SNR, since $\log(1 + x) \approx x$ for small x. Hence in low-SNR channels, diversity techniques increase the capacity about linearly, but the overall throughput is generally still poor owing to the low SNR.

Early multi-antenna systems were conservative in a manner that they valued system reliability (diversity) over aggressive data rates (spatial multiplexing). In order to get a more substantial data rate increase, the multiantenna channel should be used to send multiple independent streams, popularly known as spatial multiplexing. Specifically, the capacity can be increased as a multiple of $\min(L_t, L_r)$, i.e., capacity is limited by the minimum of the number of antennas at either the transmitter (L_t) or the receiver (L_r) [81]. Future multi-antenna systems are envisioned to have a optimal diversity-multiplexing trade-off depending on the channel condition [80].

The benefits of diversity can also be harnessed to increase the coverage area and to reduce the required transmit power, although these gains directly compete with each other, as well as with the achievable reliability and data rate. Diversity allows the received SNR at cell edges to be well above the bare minimum value required for maintaining the QoS, and opens up possibility of coverage extension without installing another BS. When radio range is not an issue, the transmit power may be lowered for a diversity enabled receiver. This in turn increases battery lifetime, allows low power cheap amplifiers, lowers interferences (CCI, ACI and ISI [82]) and reduces the probability that a hostile party will intercept the signals. Improved interference tolerance means greater ability to support additional users and therefore, higher system capacity.

Diversity systems can be also used for transmitter localization. A receiver array antenna can be used to localize the transmitter, just as we can use our both ears to localize the source of a sound in a room without using our eyes. This has application in positioning services and emergency call localization [83]. Power control algorithms also perform better with diversity as precise power control is possible.

3.3.4 Transmitter and Receiver Diversity

Traditionally diversity has been realized in the receiver, often with the help of multiple antennas attached to the receiver. Fig. 3.6 shows a generic model with L receiver antennas. Each of the channels, plus the corresponding receiver circuit, is called a branch and the outputs of the channels are processed and routed to the demodulator by a diversity combiner. The combining shown here is performed before the demodulation takes place, also known as pre-detection combining. However, the combining may also be done in a post-detection fashion, i.e. after demodulating signal from each branch. Details of pre- and post-detection combining may be found in Section 3.4.1.

In Fig. 3.6, each branch is assumed to exhibit slow flat fading which may be characterized by an attenuation factor α_k and a phase term ϕ_k, where $k \in \{1, 2, \ldots, L\}$. Thus, from (3.27), the signal in the kth branch fed to the diversity combiner may be expressed as

$$\tilde{r}_k(t) = \alpha_k \exp(-j\phi_k)\tilde{s}(t) + \tilde{n}_k(t) \tag{3.64}$$

where $\tilde{n}_k(t)$ denotes the baseband equivalent AWGN term in the kth path.

More recent work has focused on investigating how techniques such as transmit diver-

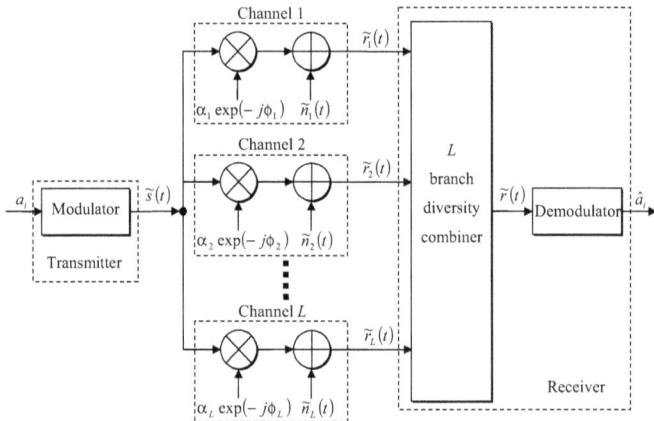

Figure 3.6: Baseband equivalent model of a diversity receiver with L diversity paths.

sity and space time codes can be applied to maximize diversity gain. Transmit diversity obviates the major disadvantage of receive diversity being the difficulty of locating multiple receive antennas far enough apart in small mobile device [84]. However, in receiver diversity the channel can be estimated (by non-blind or blind methods) since the signal has travelled through the channel before being observed at receiver. Also the interference present at the receiver input can be characterized and cancelled. On the other hand, in transmit, the channel is encountered after the signal leaves the antenna array. The use of transmit diversity therefore requires prior knowledge of the channel state information (CSI). The complete description of CSI includes channel impulse response, location information, vehicle speed, signal strength, interference level, and interference modelling [85]. It is generally assumed that CSI is available at the receiver of most antenna systems. However, CSI is not automatically available at the transmitter [86], it requires in most cases a feedback channel, which consumes system bandwidth [83] and often necessiates complex optimization [87]. The estimation and feedback of CSI takes some time, and the channel state must remain constant over that period [88]. Moreover, interference reduction in transmit diversity requires knowledge of the channels to the co-channel subscribers. Again these are difficult to estimate. Both these factors make the transmit processing challenging [82]. For transmitter diversity we also have power penalty as the total available power is divided into all antennas. For example, with two transmit antennas, a 3 dB power disadvantage occurs because the total transmit power is fixed, and therefore, each antenna must transmit 3 dB less power. Another disadvantage of space-time codes is the high decoding complexity, which grows exponentially as a function of both the required capacity and diversity order [86].

3.3.5 Macro and Micro Diversity

By selecting a base station or access point which is not shadowed when others are, a mobile receiver can improve substantially the average SNR on the forward (base to mobile) link. This is called macroscopic diversity, since the mobile is taking advantage of large separations (the macro system differences) between the serving BSs/ access points (APs). Macrodiversity attempts to mitigate the random large-scale fading effects caused by shadowing from buildings, objects and geographical terrain. As this diversity is generally implemented by combining signals received from several BS/ AP, it requires a coordination among them.

Diversity techniques that mitigate the effect of small-scale fading are called microdiversity, and that is the focus of this book. Microscopic diversity schemes use two or more uncorrelated received signals, with the same large-scale fading experienced in those signals. There are several ways in which we can generate uncorrelated fading signals at the receiver (e.g. space diversity, time diversity, frequency diversity etc.) and are described next.

3.3.6 Diversity Types

The methods by which diversity can be achieved generally fall into one of the following categories or a mixture of them.

Space Diversity

Space diversity is achieved by using multiple transmit or receive antennas separated physically by a short distance. The spatial separation between the multiple antennas is chosen so that the diversity branches experience uncorrelated fading. The separation in general varies with antenna height and with frequency. The higher the frequency, the closer the two antennas can be to each other. Typically, in a uniform scattering environment with isotropic antennas, the minimum antenna separation required at receiver is approximately $\lambda/2$, where λ denotes the wavelength of the transmitted signal. Receiver space diversity (/microdiversity) is used to mitigate the small-scale fading effects, while transmit space diversity (/macrodiversity) is used to avoid shadowing effects. Taking into account the shadowing effect, usually a separation of at least 10λ is required between two adjacent antennas at transmitter end [78]. Space diversity does not require extra system power or bandwidth, but incurs additional cost for the extra antennas.

Frequency Diversity

Frequency diversity is achieved by transmitting the same signal at different carrier frequencies, where the carriers are separated by at least the coherence bandwidth of the channel. In an urban or suburban environment the coherence bandwidth may be 300 kHz or more [78]. Thus the fading appears to be frequency non-selective to the narrowband signals used in personal communication systems (PCS), and there exists a high degree of correlation between the signals transmitted with different carrier frequencies rendering the frequency diversity to be ineffective. Apart from this, frequency diversity is neither a bandwidth efficient solution (as it tries to improve link transmission quality at the cost of extra frequency bandwidth) nor a power efficient technique (additional transmit power required to modulate the same signal at different frequency bands). However, frequency diversity may be useful for wideband signals where the fading is frequency selective. Frequency hop spread spectrum (FHSS) systems can exploit frequency diversity through the principle of fast frequency hopping, where each symbol is transmitted sequentially on multiple hops/ carriers that experience uncorrelated fading. OFDM technique may also be viewed as another example where frequency diversity is realized. In OFDM, a high data rate signal is subdivided into many parallel low data rate signals, each modulated with a different sub-carrier which are orthogonal to each other. For any particular sub carrier the frequency selective fading appears as a flat fading.

Time Diversity

Time diversity is achieved by transmitting the same signal at different times, where the time difference between the repeated transmissions is greater than the channel coherence time. Time diversity is particularly useful in mobile radio environments where the relative motion between transmitter and receiver results in large Doppler spread (or small coherence time). The signal experiences fast fading, and the low degree of correlation between successive fades may be exploited through some sort of time diversity. Time diversity does not require increased transmit power, but it does decrease the data rate since data is repeated in the diversity time slots rather than sending new data in these time slots. In addition to the redundant transmission, this diversity introduces a significant signal processing delay, especially when the channel coherence time is large. In practice, time diversity is realized through error correction coding, bit interleaving, and automatic repeat request (ARQ).

Angle Diversity

The scattering of signals from transmitter to receiver generates received signals from different directions that are uncorrelated with each other. Thus, two or more directional antennas can be pointed in different directions to achieve angle or directional diversity.

This scheme is more effective at the mobile station than at the base station since the scattering is from local buildings and vegetation and is more pronounced at street level than at the height of base station antennas. Angle diversity requires either a sufficient number of directional antennas to span all possible directions of arrival or a single antenna whose directivity can be steered to the angle of arrival of one of the multipath components (preferably the strongest one). Also the SNR may decrease due to the loss of multipath components that fall outside the receive antenna beamwidth, unless the directional gain of the antenna is sufficiently large to compensate for this lost power. Angle diversity is often realized with beamforming techniques and provides good results only when the operating frequency is ≥ 10 GHz [78]. In the past, angle diversity has been extensively used in troposcatter channels.

Polarization diversity

The horizontal and vertical polarization components transmitted by two polarized antennas at the base station and received by two polarized antennas at the mobile station can provide two uncorrelated fading signals. Since the scattering angle relative to each polarization is random, it is highly improbable that signals received on the two differently polarized antennas would be simultaneously in deep fades. While this method provides with only two diversity branches, it allows the antenna elements to be co-located and thereby realizing a compact antenna structure. Another advantage of polarization diversity is the ability to recover from a polarization mismatch, which occurs when the polarization of the transmit antenna and the receive antenna are different. Perhaps this is why this form of diversity is extensively used in UHF/ microwave communiations. However, there are several limitations of polarization diversity. First, it is viable for fixed wireless links only. Polarization diversity results in 3 dB power reduction at the transmitting site since the power must be split into two different polarized antennas. Also it is affected by an intrinsic power imbalance [89]. The nature of electromagnetic wave propagation dictates that the polarization orthogonal to the obstacle is attenuated more than the polarization parallel to the obstacle. Considering that buildings are typical obstacles in wireless channels, the horizontal polarization is expected to be attenuated more than the vertical polarization.

Path Diversity

In direct sequence spread spectrum (DSSS) CDMA systems, path diversity is obtained with a rake receiver by resolving multipath signal components. Multipath components are practically uncorrelated when their relative propagation delay $(\tau_i - \tau_j)$ exceeds one chip period (T_c). Rake receiver provides a separate correlation receiver (finger) for each of the multipath signals. Each correlator detects a time-delayed version of the original transmission, which are combined to improve the SNR at the receiver.

Some other diversity types worth mentioning include field diversity [90], antenna pattern diversity [91], Doppler diversity [92], site diversity [93], route diversity [94], relay diversity [95] etc. According to Lee [90] an E field E_z, also generates independent magnetic components H_y and H_x. By using an energy density loop antenna, it is possible to extract these three signals and achieve field diversity. Antenna pattern diversity, on the other hand, is realized with antennas with radiation patterns that have a minimal overlap [91]. Doppler spread induced by temporal channel variations can provide another means for diversity that can be exploited to combat fading [92]. Wireless relays are now being introduced for capacity enhancement of cellular systems and such relay assisted communication systems are shown to achieve the benefits of spatial diversity without requiring physical antenna arrays [95]. Further, there also exists many hybrid diversity techniques. For example, space and time diversity can be combined together by using space-time coding (STC) techniques. Similarly space-frequency coding (SFC) and space-time-frequency coding (STFC) are also possible. It may be noted that many authors prefer to include angle and polarization diversity under the heading space diversity, as multiple antennas are required for all of them. Also, rake reception is sometimes categorized separately as implicit diversity, because no redundant signal transmission is employed like other explicit diversity techniques (time, frequency etc.). Finally, though we dealt with many types of diversity, we will focus on space diversity throughout this book. One main advantage of space diversity relative to time and frequency diversity is that no additional bandwidth or power is needed in order to take advantage of spatial diversity. The cost of each additional antenna, its RF chain, and the associated signal processing required to modulate or demodulate multiple spatial streams may not be negligible, but this trade-off is often very attractive for a small number of antennas. In the rest of the book, space diversity serves as a reference to describe the different combining techniques, although the combining techniques can be applied to any type of diversity.

3.4   Diversity Combining

The performance of any diversity system depends on the combining technique used to merge the signals received on the disparate diversity branches. The output signals from diversity antennas can be selected or combined in several ways to optimize the received signal power. There are different possible combining methods employed in receivers, among which the four widely considered techniques are selection combining (SC), maximal ratio combining (MRC), equal gain combining (EGC) and switched combining (SWC). A hybrid of these techniques (e.g. generalized selection combining) are also sometimes used.

3.4.1 Pre and Post Detection Combining

For pre-detection combining, received signals are combined first and then the combiner output is demodulated. In case of post-detection, all received signals are demodulated and then combined. In other words, diversity combining that takes place at RF is called pre-detection combining, while diversity combining that takes place at baseband is called post-detection combining. Post-detection combining outperforms pre-detection combining in general, and the receiver is usually much simpler to implement. On the other hand, for pre-detection combining, each received signal is co-phased at the RF. So a more complicated receiver structure is required due to in-phase addition. However, for post-detection combining, multiple RF demodulator chains (correlators or matched filters) are required which makes the receiver bulky and power-hungry. SC, MRC, EGC, and SWC techniques can be used with both pre-detection combining and post-detection combining. As far as this book is concerned, our study is limited to pre-detection diversity combining only.

3.4.2 Linear Combining

A linear diversity combiner consisting of L antennas is shown in Fig. 3.7. Each of these antennas realizes one diversity branch and provides differently faded replicas of the same information bearing signal. Further, it is assumed that the fading is slow, flat, independent and identically distributed (IID), i.e. the fading processes are uncorrelated and all channels have the same mean signal power. The assumption of equal mean SNRs at the branches is very reasonable in view of the fact that the array length, in practice, is very small compared with the distance from the transmitter so that all elements experience approximately the same path loss and shadowing loss.

According to the given diagram, the equivalent baseband signal $r_k(t)$ received on each antenna is weighted by a complex weight factor w_k and summed to produce the baseband signal $r(t)$ at combiner output. Mathematically,

$$\tilde{r}(t) = \sum_{k=1}^{L} w_k \tilde{r}_k(t) \tag{3.65}$$

Using the complex envelope of the received signal at the kth diversity branch given in (3.64), we find that the corresponding complex envelope of the linear combiner output given in (3.65) may also be written as

$$\tilde{r}(t) = \tilde{s}(t) \underbrace{\sum_{k=1}^{L} w_k \alpha_k \exp(-j\phi_k)}_{\text{signal term}} + \underbrace{\sum_{k=1}^{L} w_k \tilde{n}_k(t)}_{\text{noise term}} \tag{3.66}$$

81

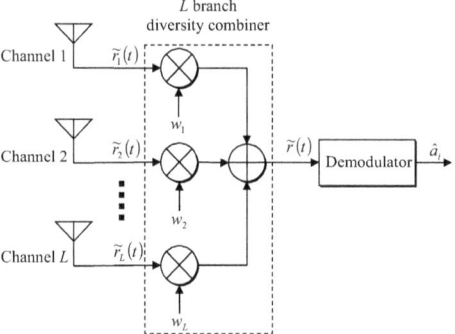

Figure 3.7: Generalized model for pre-detection linear combiner.

where $\{\tilde{n}_k(t)\}_{k=1}^{L}$ represents mutually independent noise processes, which are uncorrelated with the fading process.

There are basically three performance parameters, namely the outage probability, average SNR at combiner output, and the average BEP/SEP to denote the performance of a diversity combiner. Outage probability is defined as the probability that the instantaneous SNR (γ_{LC}) at combiner output is below some target SNR value (γ_o), and is given by

$$P_{o,LC} = \Pr[\gamma_{LC} \leq \gamma_o] = \int_0^{\gamma_o} f_{\gamma_{LC}}(\gamma)d\gamma = F_{\gamma_{LC}}(\gamma_o) \tag{3.67}$$

where $f_{\gamma_{LC}}(\gamma)$ and $F_{\gamma_{LC}}(\gamma)$, respectively, denote the PDF and CDF of γ_{LC}. Eq. (3.67) also shows the fact that once the CDF of γ_{LC} is known, the outage probability can be calculated by simply replacing γ with γ_o, the target SNR. The CDF $F_{\gamma_{LC}}(\gamma)$ depends on the fading environment. In the rest of this subsection we would be considering Rayleigh fading to obtain numerical values for the outage probability.

The improvement in average SNR is generally defined in terms of the gain in SNR (in dB) as

$$G_{\text{SNR}} = 10 \log_{10} \left(\frac{\bar{\gamma}_{LC}}{\bar{\gamma}} \right) \tag{3.68}$$

assuming $\bar{\gamma}_k = \bar{\gamma}; \forall k$, i.e. identical fading over all the L diversity branches.

The average error probability for a given modulation scheme is determined by averaging the CEP $P_e(\gamma)$ over $f_{\gamma_{LC}}(\gamma)$ as demonstrated in (3.32). The CEP $P_e(\gamma)$ is a modulation dependent quantity, and is derived in Chapter 1.3 for all different modulations of interest.

For comparing the improvement in error performance with different combining methods, we consider simple BPSK modulation. In a Rayleigh fading channel its BEP is given by [27]

$$
\begin{aligned}
P_e &= \int_0^\infty P_e(\gamma) f_\gamma(\gamma) d\gamma \\
&= \frac{1}{2\bar{\gamma}} \int_0^\infty \mathrm{erfc}(\sqrt{\gamma}) \exp\left(-\frac{\gamma}{\bar{\gamma}}\right) d\gamma \\
&= \frac{1}{2}\left(1 - \sqrt{\frac{\bar{\gamma}}{1+\bar{\gamma}}}\right) \\
&\approx \frac{1}{4\bar{\gamma}} \qquad ; \text{for large } \bar{\gamma}
\end{aligned}
\tag{3.69}
$$

The improvement in error performance is often denoted with the diversity order. A diversity system is said to have diversity order L if, in Rayleigh fading,

$$
L = \lim_{\bar{\gamma} \to \infty}\left(-\frac{\log P_e}{\log \bar{\gamma}}\right)
\tag{3.70}
$$

or equivalently

$$
P_e \propto \frac{1}{\bar{\gamma}^L}
\tag{3.71}
$$

for large $\bar{\gamma}$. In a diversity system, therefore, we expect the BER to be a linear function of the SNR (in a log-log plot) having a slope equal to the diversity order.

3.4.3 SC

Selection combining (SC) is the simplest type of combiner, in that it simply estimates the instantaneous signal strengths of each of the L branches and selects the highest one. Mathematically, as the branch with the highest instantaneous SNR is chosen, we state the weight condition as [96]

$$
w_k = \begin{cases} 1 & ; \gamma_k = \max(\gamma_1, \gamma_2, \cdots, \gamma_L) \\ 0 & ; \text{otherwise} \end{cases}
\tag{3.72}
$$

Since the element chosen is the one with the maximum SNR, the output SNR of the selection diversity scheme is

$$
\gamma_{SC} = \max_k\{\gamma_k\} \qquad ; k = 1, 2, \cdots, L
\tag{3.73}
$$

Since the useful energy on the other branches (other than the one that is selected) are ignored, SC is clearly suboptimal, but its simplicity and reduced hardware requirements (selection combiner would need no more than a comparator and a fast switch, phase shifters or variable gains are not required) make it attractive in many cases.

Let us now investigate the probability of outage, BEP, and resulting improvement in average SNR with SC. To derive $P_{o,SC}$, we note that the SC combiner output SNR would drop below the target SNR γ_o when the SNR of all elements is below the the target SNR. Assuming IID Rayleigh fading,

$$
\begin{aligned}
P_{o,SC} &= \Pr[\gamma_{SC} \leq \gamma_o] \\
&= \Pr[(\gamma_1 \leq \gamma_o) \cap \cdots \cap (\gamma_L \leq \gamma_o)] \\
&= \prod_{k=1}^{L} \Pr[\gamma_k \leq \gamma_o] \\
&= \prod_{k=1}^{L} F_{\gamma_k}(\gamma_o) \\
&= \left[1 - \exp\left(-\frac{\gamma_o}{\bar{\gamma}} \right) \right]^L
\end{aligned}
\tag{3.74}
$$

The outage probability therefore decreases exponentially with the number of elements. Fig. 3.8 illustrates the improvement in outage probability as a function of the number of elements in the array. As is clearly seen, going from $L = 1$ to $L = 2$ at 1% outage probability there is an approximate 12 dB reduction in required SNR, and at 0.01% outage probability there is an approximate 20 dB reduction in required SNR. However, at 0.01% outage, going from two branch to three-branch diversity results in an additional reduction of approximately 7 dB, and from three-branch to four-branch results in an additional reduction of approximately 4 dB. Clearly the power savings is most substantial going from no diversity to two-branch diversity, with diminishing returns as the number of branches is increased [4].

$P_{o,SC}$ also represents the CDF of the output SNR as a function of the threshold γ_o. The CDF and PDF of the output SNR, $f_{\gamma,SC}$, is therefore

$$
F_{\gamma_{SC}}(\gamma) = \left[1 - \exp\left(-\frac{\gamma}{\bar{\gamma}} \right) \right]^L
\tag{3.75}
$$

and

$$
f_{\gamma_{SC}}(\gamma) = \frac{dF_{\gamma_{SC}}(\gamma)}{d\gamma} = \frac{L}{\bar{\gamma}} \exp\left(-\frac{\gamma}{\bar{\gamma}} \right) \left[1 - \exp\left(-\frac{\gamma}{\bar{\gamma}} \right) \right]^{L-1}
\tag{3.76}
$$

respectively.

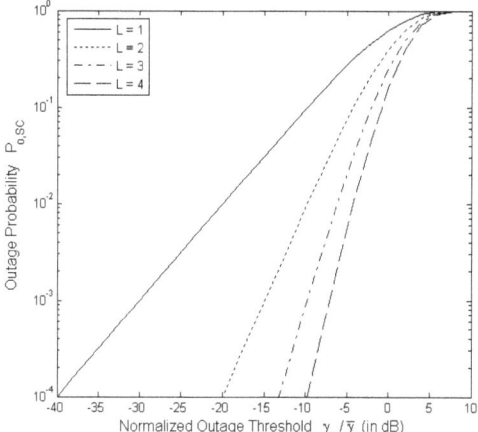

Figure 3.8: Outage probability with SC in Rayleigh fading.

The average SNR at the selection combiner output is given by [96]

$$
\begin{aligned}
\bar{\gamma}_{SC} &= \int_0^\infty \gamma f_{\gamma_{SC}}(\gamma)d\gamma = \bar{\gamma}\sum_{k=1}^{L}\frac{1}{k} \\
&\approx \bar{\gamma}\left(C + \ln L + \frac{1}{2L}\right) \quad ; \qquad \text{for} \quad L \geq 3
\end{aligned}
\tag{3.77}
$$

where $C \approx 0.577215$ is the Euler's constant. Hence the improvement in SNR over that of a single element is of order of $\ln L$.

The BEP of BPSK with SC is given by [97, (5-193)]

$$
\begin{aligned}
P_e &= \int_0^\infty P_e(\gamma)f_{\gamma_{SC}}(\gamma)d\gamma \\
&= \frac{L}{2\bar{\gamma}}\int_0^\infty \mathrm{erfc}(\sqrt{\gamma})\exp\left(-\frac{\gamma}{\bar{\gamma}}\right)\left[1 - \exp\left(-\frac{\gamma}{\bar{\gamma}}\right)\right]^{L-1} \\
&= \frac{1}{2}\sum_{k=0}^{L-1}\binom{L}{k+1}(-1)^k\left(1 - \sqrt{\frac{\bar{\gamma}}{1+k+\bar{\gamma}}}\right)
\end{aligned}
\tag{3.78}
$$

3.4.4   MRC

In absence of interference, maximal ratio combining (MRC) serves as the optimal combining scheme. In MRC, the complex weighting parameters (w_k) are adjusted at every signalling instant to maximize the output SNR. The optimization process starts with

85

finding the output SNR of the linear combiner [98]

$$\gamma_{LC} = \frac{\left|\tilde{s}(t)\sum_{k=1}^{L}w_k\alpha_k\exp(-j\phi_k)\right|^2}{\left|\sum_{k=1}^{L}w_k\tilde{n}_k(t)\right|^2}$$

$$= \frac{|\tilde{s}(t)|^2}{|\tilde{n}_k(t)|^2}\frac{\left|\sum_{k=1}^{L}w_k\alpha_k\exp(-j\phi_k)\right|^2}{\sum_{k=1}^{L}|w_k|^2} \qquad (3.79)$$

$$= \left(\frac{E_s}{N_0}\right)\frac{\left|\sum_{k=1}^{L}w_k\alpha_k\exp(-j\phi_k)\right|^2}{\sum_{k=1}^{L}|w_k|^2}$$

The requirement is to maximize γ_{LC} with respect to the $\{w_k\}_{k=1}^{L}$. The task may be performed easily with the Cauchy-Schwarz inequality. According to this inequality if $\{a_k\}_{k=1}^{L}$ and $\{b_k\}_{k=1}^{L}$ denote sets of complex numbers, then

$$\left|\sum_{k=1}^{L}a_kb_k\right|^2 \le \sum_{k=1}^{L}|a_k|^2\sum_{k=1}^{L}|b_k|^2 \qquad (3.80)$$

which holds with equality for $a_k = cb_k^*$, where c is some arbitrary complex constant. Applying (3.80) in the numerator term in the final expression of (3.79) we obtain

$$\gamma_{LC} \le \left(\frac{E_s}{N_0}\right)\sum_{k=1}^{L}|\alpha_k\exp(-j\phi_k)|^2 \le \left(\frac{E_s}{N_0}\right)\sum_{k=1}^{L}\alpha_k^2 \qquad (3.81)$$

Let the complex constant is equal to unity ($c = 1$). In that case, the equality in (3.81) holds for the weight condition

$$w_k = [\alpha_k\exp(-j\phi_k)]^* = \alpha_k^*\exp(j\phi_k) \qquad (3.82)$$

giving the optimal weighting factors. From (3.82) one can find that the weighting factor w_k for the kth diversity branch has a magnitude proportional to the amplitude α_k of the signal and a phase that cancels the signal phase ϕ_k. The phase alignment permits fully coherent addition of the L branches by the MRC combiner. In some sense, this answer is expected since the solution is effectively the matched filter for the fading signal and we know that the matched filter is optimal in the single user case.

Eq. (3.81) with the equality sign defines the instantaneous output SNR of the MRC combiner

$$\gamma_{MRC} = \left(\frac{E_s}{N_0}\right)\sum_{k=1}^{L}\alpha_k^2 = \sum_{k=1}^{L}\gamma_k \qquad (3.83)$$

The output SNR is, therefore, the sum of the SNR at each element. Assuming the equal mean signal power in all the branches and from (3.83), the expected value of the output SNR is therefore L times the average SNR at each element, i.e.,

$$\bar{\gamma}_{MRC} = \sum_{k=1}^{L} \bar{\gamma}_k = L\bar{\gamma} \tag{3.84}$$

which indicates that on average, the SNR improves by a factor of L. This is significantly better than the factor of $\ln L$ improvement in the selection diversity case.

According to (3.84), for a slow flat Rayleigh fading channel, γ_{MRC} is equal to the sum of L exponentially distributed RVs. From probability theory, the PDF of such a sum is known to be chi-square with $2L$ degrees of freedom

$$f_{\gamma_{MRC}}(\gamma) = \frac{1}{(L-1)!} \frac{\gamma^{L-1}}{\bar{\gamma}^L} \exp\left(-\frac{\gamma}{\bar{\gamma}}\right) \tag{3.85}$$

Using this PDF, the outage probability for a threshold γ_o is [4]

$$
\begin{aligned}
P_{o,MRC} &= \Pr[\gamma_{MRC} \leq \gamma_o] = \frac{1}{(L-1)!\bar{\gamma}^L} \int_0^{\gamma_o} \gamma^{L-1} \exp\left(-\frac{\gamma}{\bar{\gamma}}\right) d\gamma \\
&= 1 - \exp\left(-\frac{\gamma_o}{\bar{\gamma}}\right) \sum_{k=0}^{L-1} \left(\frac{\gamma_o}{\bar{\gamma}}\right)^k \frac{1}{k!}
\end{aligned}
\tag{3.86}
$$

Fig. 3.9 illustrates the performance of a MRC with multiple antenna elements. Again, the large performance gains from using two or more elements is clear. When compared to SC we find that at outage probability of 1%, MRC is about 3dB better than SC for $L = 2$.

The final figure of merit is the BER in a BPSK system. The BER is given by [75]

$$
\begin{aligned}
P_e &= \int_0^\infty P_e(\gamma) f_{\gamma_{MRC}}(\gamma) d\gamma \\
&= \frac{1}{2} \frac{1}{(L-1)!\bar{\gamma}^L} \int_0^\infty \text{erfc}(\sqrt{\gamma})\gamma^{L-1} \exp\left(-\frac{\gamma}{\bar{\gamma}}\right) d\gamma \\
&= \frac{1}{(L-1)!} \left(\frac{1-\mu}{2}\right)^L \sum_{k=0}^{L-1} \frac{(L-1+k)!}{k!} \left(\frac{1+\mu}{2}\right)^k \\
&\approx \binom{2L-1}{L} \left(\frac{1}{4\bar{\gamma}}\right)^L \quad ; \quad \text{for large} \quad L
\end{aligned}
\tag{3.87}
$$

where $\mu = \sqrt{\bar{\gamma}/(1+\bar{\gamma})}$ and the approximation for large value of L is obtained using only the final term of the summation. Note the BER reduces exponentially as a function of L.

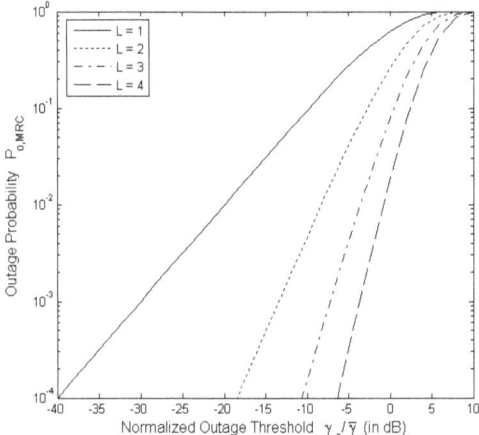

Figure 3.9: Outage probability with MRC in Rayleigh fading.

3.4.5   EGC

In Section 3.4.4 we developed a combiner that is optimal in the sense of SNR. However, the technique requires the weights to vary with the fading signals, the magnitude of which may fluctuate over several tens of dB. The equal gain combiner (EGC) sidesteps this problem by setting unit gain at each element. In EGC

$$w_k = \exp(j\phi_k) \qquad ; k = 1, 2, \cdots, L \tag{3.88}$$

and the corresponding complex envelope of the linear combiner output is defined by

$$\tilde{r}(t) = \underbrace{\tilde{s}(t) \sum_{k=1}^{L} \alpha_k}_{\text{signal term}} + \underbrace{\sum_{k=1}^{L} \tilde{n}_k(t) \exp(-j\phi_k)}_{\text{noise term}} \tag{3.89}$$

and the output SNR of the linear combiner is [75]

$$
\begin{aligned}
\gamma_{EGC} &= \frac{\left| \tilde{s}(t) \sum_{k=1}^{L} \alpha_k \right|^2}{\left| \sum_{k=1}^{L} \tilde{n}_k(t) \exp(-j\phi_k) \right|^2} \\
&= \left(\frac{E_s}{LN_0} \right) \left| \sum_{k=1}^{L} \alpha_k \right|^2
\end{aligned}
\tag{3.90}
$$

with the average SNR being

$$\bar{\gamma}_{EGC} = \left(\frac{E_s}{LN_0}\right) E\left\{\left|\sum_{k=1}^{L} \alpha_k\right|^2\right\} = \left(\frac{E_s}{LN_0}\right) \sum_{i=1}^{L}\sum_{j=1}^{L} E\{\alpha_i\alpha_j\} \tag{3.91}$$

With Rayleigh fading, $E\{\alpha_k^2\} = 2\sigma^2$ and $E\{\alpha_k\} = \sqrt{\pi/2}\sigma$. Furthermore, if the branches experience uncorrelated fading, then $E\{\alpha_i\alpha_j\} = E\{\alpha_i\}E\{\alpha_j\}$ for $i \neq j$. Hence,

$$\begin{aligned}
\bar{\gamma}_{EGC} &= \left(\frac{E_s}{LN_0}\right)\left[2L\sigma^2 + L(L-1)\frac{\pi\sigma^2}{2}\right] \\
&= \left(\frac{2\sigma^2 E_s}{N_0}\right)\left[1 + (L-1)\frac{\pi}{4}\right] \\
&= \bar{\gamma}\left[1 + (L-1)\frac{\pi}{4}\right]
\end{aligned} \tag{3.92}$$

Eq. (3.92) depicts that, despite being significantly simpler to implement, EGC results in an improvement in SNR that is comparable to that of the optimal MRC. The SNR of both combiners increases linearly with L.

In Rayleigh fading channel, the CDF and PDF for γ_{EGC} does not exist in closed form for $L > 2$. However, for $L = 2$ and $\bar{\gamma}_1 = \bar{\gamma}_2 = \bar{\gamma}$, the CDF is equal to [75]

$$\begin{aligned}
F_{\gamma_{EGC}}(\gamma) &= 1 - \exp\left(-\frac{2\gamma}{\bar{\gamma}}\right) \\
&\quad - \sqrt{\frac{\pi\gamma}{\bar{\gamma}}}\exp\left(-\frac{\gamma}{\bar{\gamma}}\right)\left[1 - 2Q\left(\sqrt{\frac{2\gamma}{\bar{\gamma}}}\right)\right]
\end{aligned} \tag{3.93}$$

Hence the outage probaility would be

$$\begin{aligned}
P_{o,EGC} &= 1 - \exp\left(-\frac{2\gamma_o}{\bar{\gamma}}\right) \\
&\quad - \sqrt{\frac{\pi\gamma_o}{\bar{\gamma}}}\exp\left(-\frac{\gamma_o}{\bar{\gamma}}\right)\left[1 - 2Q\left(\sqrt{\frac{2\gamma_o}{\bar{\gamma}}}\right)\right]
\end{aligned} \tag{3.94}$$

Fig. 3.10 shows the corresponding outage probability. The outage probability with no diversity is also shown for comparison purpose. Differentiating (3.93) we get the corresponding PDF as [75]

$$\begin{aligned}
f_{\gamma_{EGC}}(\gamma) &= \frac{1}{\bar{\gamma}}\exp\left(-\frac{2\gamma}{\bar{\gamma}}\right) - \sqrt{\frac{\pi\gamma}{\bar{\gamma}}}\left(\frac{1}{2\gamma} - \frac{1}{\bar{\gamma}}\right)\exp\left(-\frac{\gamma}{\bar{\gamma}}\right) \\
&\quad \times \left[1 - 2Q\left(\sqrt{\frac{2\gamma}{\bar{\gamma}}}\right)\right]
\end{aligned} \tag{3.95}$$

The BER of BPSK with dual branch EGC can be obtained by using the PDF in (3.95)

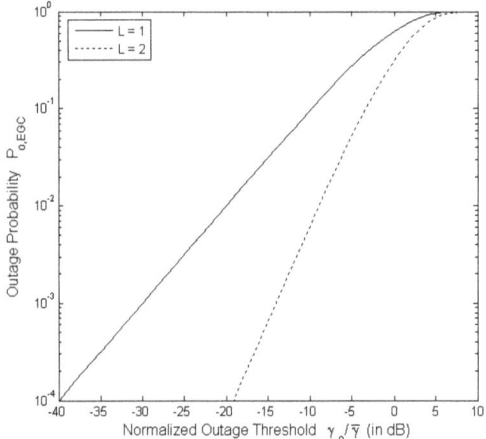

Figure 3.10: Outage probability with EGC in Rayleigh fading.

[75]

$$
\begin{aligned}
P_e &= \int_0^\infty P_e(\gamma) f_{\gamma_{EGC}}(\gamma) d\gamma \\
&= \frac{1}{2} \int_0^\infty \mathrm{erfc}(\sqrt{\gamma}) \left[\frac{1}{\bar{\gamma}} \exp\left(-\frac{2\gamma}{\bar{\gamma}} \right) \right. \\
&\quad \left. - \sqrt{\frac{\pi\gamma}{\bar{\gamma}}} \left(\frac{1}{2\gamma} - \frac{1}{\bar{\gamma}} \right) \exp\left(-\frac{\gamma}{\bar{\gamma}} \right) \left\{ 1 - 2Q\left(\sqrt{\frac{2\gamma}{\bar{\gamma}}} \right) \right\} \right] d\gamma \\
&= \frac{1}{2} \left(1 - \sqrt{1 - \nu^2} \right)
\end{aligned} \tag{3.96}
$$

where $\nu = 1/(1 + \bar{\gamma})$.

3.4.6 SWC

A switched combiner (SWC), also known as feedback or scanning diversity [99], scans through the diversity branches until it finds a branch that has a SNR exceeding a specified threshold, meeting the QoS requirements. This diversity branch is selected and used until the SNR again drops below the threshold, and then the process repeats. Implementation of SWC is much easier than other combining techniques as it avoids continuous monitoring of all the branches.

There are mainly two variants of switched diversity, the switch and stay combining (SSC) used in conjunction with a dual branch diversity system, and switch and examine combining (SEC) for the general L branch scenario. Although initial research work was primarily focused on dual branch SSC, more recent applications have motivated studies of multibranch SEC. In this subsection, we would first study the performance of a dual-branch SSC systems and then have a quick look into SEC systems.

SSC

With SSC, the receiver switches to, and stays with, the alternate branch when the SNR drops below a specified threshold. The switching action is performed regardless of whether or not the SNR with the alternate branch is above or below the threshold. Let the SNRs associated with the two branches be denoted by γ_1 and γ_2, and let the switching threshold be denoted by γ_T. Now, if γ_{SSC} denotes the SNR per symbol of the SSC combiner output, the CDF of γ_{SSC} can be written in the following manner [73]

$$F_{\gamma_{ssc}}(\gamma) = \begin{cases} \Pr[(\gamma_1 \leq \gamma_T) \cap (\gamma_2 \leq \gamma)] & ; \gamma < \gamma_T \\ \Pr[(\gamma_T \leq \gamma_1 \leq \gamma) \cup \{(\gamma_1 \leq \gamma_T) \cap (\gamma_2 \leq \gamma)\}] & ; \gamma \geq \gamma_T \end{cases} \tag{3.97}$$

which can be expressed in terms of the CDF of the individual branches $F_\gamma(\gamma)$ as

$$F_{\gamma_{ssc}}(\gamma) = \begin{cases} F_\gamma(\gamma_T)F_\gamma(\gamma) & ; \quad \gamma < \gamma_T \\ F_\gamma(\gamma) - F_\gamma(\gamma_T) + F_\gamma(\gamma)F_\gamma(\gamma_T) & ; \quad \gamma \geq \gamma_T \end{cases} \tag{3.98}$$

Further, differentiating $F_{\gamma_{ssc}}(\gamma)$ with respect to γ, we get the PDF of the SSC output in terms of the CDF $F_\gamma(\gamma)$ and the PDF $f_\gamma(\gamma)$ of the individual branches as

$$f_{\gamma_{ssc}}(\gamma) = \frac{dF_{\gamma_{ssc}}(\gamma)}{d\gamma} = \begin{cases} f_\gamma(\gamma)F_\gamma(\gamma_T) & ; \quad \gamma < \gamma_T \\ f_\gamma(\gamma)[1 + F_\gamma(\gamma_T)] & ; \quad \gamma \geq \gamma_T \end{cases} \tag{3.99}$$

Inserting (3.39) and (3.40) in (3.98) and (3.99), one may obtain the expressions for CDF and PDF for the Rayleigh fading channel

$$F_{\gamma_{ssc}}(\gamma) = \begin{cases} \left[1 - \exp\left(-\frac{\gamma}{\bar{\gamma}}\right)\right]\left[1 - \exp\left(-\frac{\gamma_T}{\bar{\gamma}}\right)\right] & ; \gamma < \gamma_T \\ 1 - 2\exp\left(-\frac{\gamma}{\bar{\gamma}}\right) + \exp\left\{-\frac{(\gamma + \gamma_T)}{\bar{\gamma}}\right\} & ; \gamma \geq \gamma_T \end{cases} \tag{3.100}$$

$$f_{\gamma_{ssc}}(\gamma) = \begin{cases} \frac{1}{\bar{\gamma}}\exp\left(-\frac{\gamma}{\bar{\gamma}}\right)\left[1 - \exp\left(-\frac{\gamma_T}{\bar{\gamma}}\right)\right] & ; \gamma < \gamma_T \\ \frac{1}{\bar{\gamma}}\exp\left(-\frac{\gamma}{\bar{\gamma}}\right)\left[2 - \exp\left(-\frac{\gamma_T}{\bar{\gamma}}\right)\right] & ; \gamma \geq \gamma_T \end{cases} \tag{3.101}$$

The average SNR at the SSC output can be obtained by averaging γ over $f_{\gamma_{ssc}}(\gamma)$ as given by (3.99), yielding

$$
\begin{aligned}
\bar{\gamma}_{SSC} &= F_\gamma(\gamma_T) \int_0^\infty \gamma f_\gamma(\gamma) d\gamma + \int_{\gamma_T}^\infty \gamma f_\gamma(\gamma) d\gamma \\
&= \bar{\gamma} + \int_{\gamma_T}^\infty \gamma f_\gamma(\gamma) d\gamma
\end{aligned}
\tag{3.102}
$$

Differentiating (3.102) with respect to γ_T and setting the result to zero, we can easily show that $\bar{\gamma}_{SSC}$ is maximized when the switching threshold is set to $\gamma_T^* = \bar{\gamma}$. For Rayleigh fading, substituting the CDF and PDFs given in (3.40) and (3.39), we get [73]

$$
\bar{\gamma}_{SSC} = \bar{\gamma} \left[1 + \frac{\gamma_T}{\bar{\gamma}} \exp\left(-\frac{\gamma_T}{\bar{\gamma}} \right) \right]
\tag{3.103}
$$

which reduces for the optimal threshold case to

$$
\bar{\gamma}_{SSC}^* = \bar{\gamma} \left(1 + \frac{1}{e} \right)
\tag{3.104}
$$

The outage probability for SSC, $P_{o,SSC}$, can be obtained by replacing γ with γ_o in the CDF expressions given in (3.100). In this connection, we want to state that the outage probability of dual-branch SC systems is given by

$$
P_{o,SC} = [F_\gamma(\gamma_o)]^2
\tag{3.105}
$$

Note that if we substitute γ_T for γ_o in (3.100), then $P_{o,SSC}$ becomes identical with $P_{o,SC}$. Since SC can be viewed as an optimal implementation of any switched diversity system, we can conclude that the optimal switching threshold in the minimum outage probability sense is given by $\gamma_T^* = \gamma_o$. Hence, we can write the outage probability of SC and SSC systems with optimal switching thresholds as

$$
P_{o,SSC} = P_{o,SC} = \left[1 - \exp\left(-\frac{\gamma_o}{\bar{\gamma}} \right) \right]^2
\tag{3.106}
$$

A plot of $P_{o,SSC}$ is depicted in Fig. 3.11 for different switching thresholds. The nature of the curve changes at the corresponding switching threshold. The optimal outage probability (which is equal to $P_{o,SC}$) is also shown.

The setting of the predetermined threshold is an additional important system design issue for SSC. For instance, if this threshold level is chosen too high, the switching unit is almost continually switching between the two antennas, which results not only in a poor diversity gain but also in an undesirable increase in the rate of the switching transients on the transmitted data stream. On the other hand, if this threshold level is chosen too

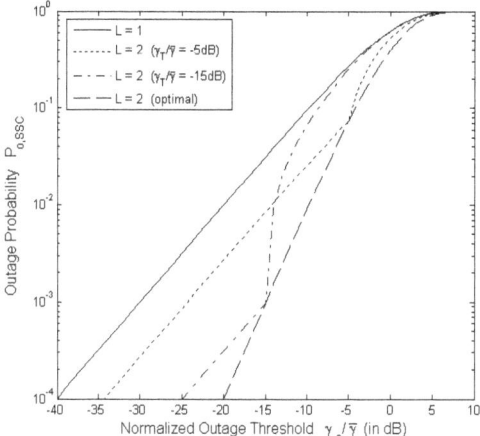

Figure 3.11: Outage probability with SSC in Rayleigh fading for various values of normalized switching threshold $\gamma_T/\bar{\gamma}$.

low, the switching unit is almost locked to one of the diversity branches even when the SNR level is quite low, and again little diversity gain is achieved.

SEC

SSC diversity systems are unable to achieve additional diversity gain when more than two diversity branches are available, as only two paths are involved at most in the diversity combining decision of SSC schemes. In this case, one should rather implement an SEC type of combining for which it is assumed that if the current path is not of acceptable quality, then the combiner switches and examines the quality of the next available path. This switching and examining process is repeated until either an acceptable path is found or all available diversity paths have been examined. In the latter case, the combiner either settles on the last examined path or connects to the receiver the path with the best quality among all examined paths.

For independent and identical diversity branches the CDF of the output SNR with L-branch SEC is given by [73]

$$
F_{\gamma_{SEC}}(\gamma) = \begin{cases} [F_\gamma(\gamma_T)]^{L-1} F_\gamma(\gamma) & ; \gamma < \gamma_T \\ [F_\gamma(\gamma) - F_\gamma(\gamma_T)] \displaystyle\sum_{k=0}^{L-1} [F_\gamma(\gamma_T)]^k + [F_\gamma(\gamma_T)]^L & ; \gamma \geq \gamma_T \end{cases} \tag{3.107}
$$

and differentiating (3.107) with respect to γ , we obtain the PDF of the output SNR as

$$f_{\gamma_{SEC}}(\gamma) = \begin{cases} f_{\gamma}(\gamma)[F_{\gamma}(\gamma_T)]^{L-1} & ;\gamma < \gamma_T \\ \\ f_{\gamma}(\gamma)\sum_{k=0}^{L-1}[F_{\gamma}(\gamma_T)]^k & ;\gamma \geq \gamma_T \end{cases} \tag{3.108}$$

Although it is difficult to find a closed-form expression for the optimal average SNR for SEC diversity systems, the outage probability may be easily calculated. Using a logic similar to the SSC case, we find that the minimum outage probability with general L branch SEC would be

$$P_{o,SEC} = P_{o,SC} = \left[1 - \exp\left(-\frac{\gamma_o}{\bar{\gamma}}\right)\right]^L \tag{3.109}$$

for Rayleigh fading channel.

3.4.7 Applicability of Different Combining Schemes

In the previous subsections, we have discussed the four major combining methodologies namely, SC, MRC, EGS and SWC. In SC, the receiver chooses the path with the highest average SNR and performs detection based on the signal from the selected path. Usually the signal strength is taken as a measure of signal quality, but other measures can be used too. EGC seeks to improve on this by co-phasing the signals and adding them together. MRC is the optimum method where signals are co-phased and weighted by their SNRs before combining. The fourth method, SWC, is an suboptimal version of SC. From the decision strategies itself, we may conclude that the performance of these diversity schemes would have the following order, SWC<SC<EGC<MRC, which may be further confirmed from the improvement in the outage probability as shown for individual combining schemes in Fig. 3.8 - Fig. 3.11. In this subsection, we describe the comparison between these four combining methods on the basis of average SNR improvement and error probability reduction, illustrated in Fig. 3.12 and Fig. 3.13 respectively.

Both these figures show that SC is inferior to EGC or MRC. However, MRC and EGC need some of the channel state information (CSI) from all the received signals and require dedicated RF chain for each diversity branch. SC dispenses with the need of multiple RF chains, as only one RF chain at a time needs to be activated. Further, if the demodulator uses a noncoherent or differential detection algorithm, i.e. the receiver does not come with an in-built synchronization circuitry, SC is an ideal match. Implementation of MRC or EGC would require extra co-phasor circuit blocks which may be avoided only when the demodulation is of coherent type. Also, when noises or interference signals are correlated, selection/ switched diversity may become more competitive. Switch diversity simplifies

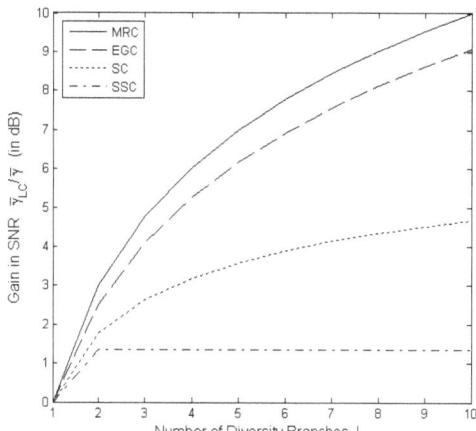

Figure 3.12: Improvement in average SNR for different combining techniques in Rayleigh fading channel.

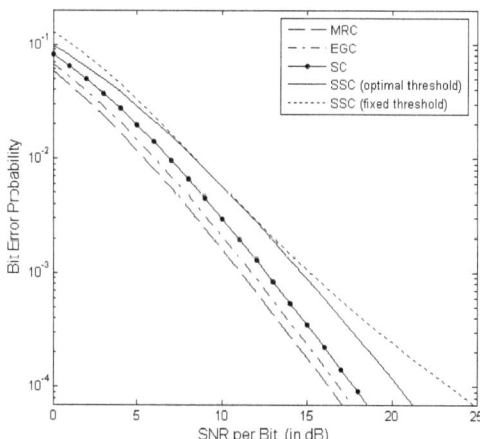

Figure 3.13: Comparison of BEP for different combining techniques in Rayleigh fading channel.

the design further, but the receiver output becomes discontinuous at the switching instants, causing unwanted amplitude and phase variations. Also, one has to keep in mind that in a fading environment, selection/ switched diversity is sensible only if the selection/

switching rate is much faster than the fading rate.

When the combiner has accurate knowledge of CSI, MRC performs better than EGC, particularly when the noise variance is high and the number of diversity branches are relatively high. For an interference-limited cellular system, MRC would be strongly preferred to either EGC or SC, despite the fact that the latter techniques are somewhat simpler. An additional important advantage of MRC in frequency-selective fading channels is that all the frequency diversity can be used, whereas an RF antenna-selection algorithm would simply select the best average antenna and then must live with the potentially deep fades at certain frequencies. However, it should be noted that although MRC does in fact maximize SNR and generally performs well, it may not be optimal in many cases since it ignores interference powers, the statistics of which may differ from branch to branch.

Since knowledge of channel fading amplitudes is needed for MRC, this scheme can be used in conjunction with unequal energy signals, such as MQAM or any other amplitude/phase modulations, while EGC is often limited in practice to coherent modulations with equal energy symbols (e.g. MPSK). Indeed, for signals with unequal energy symbols such as MQAM, estimation of the path amplitudes is needed anyway for automatic gain control (AGC) purposes, and thus for these modulations MRC should be used to achieve better performance.

Chapter Summary

Wireless transmission of modulated digital signals suffer severe degradation due to the multipath and time-varying nature of the channel. This phenomenon is known as fading and can be mitigated through the exploitation of spatial diversity combining. This chapter describes the mobile wireless channel, signal degradations due to different fading mechanisms, and various diversity techniques. The impulse response of a multipath time-varying channel is developed, and it was shown that depending on the propagation environment, the attenuation profile of the generic channel model obeys different statistical distributions (e.g., Rayleigh, Rician etc.). In the subsequent chapters, these statistical models will be used to assess the performance of different modulation schemes, described in Chapter 1.3. Various means of achieving diversity are discussed next, highlighting receiver space diversity, and pointing out its merits over others. The performance of any diversity system depends on the combining technique used to merge the signals received from the disparate diversity branches, and there exists a lot of such techniques. In the current chapter, four basic combining methods and their compatibility with coherent and non-coherent modulations are discussed.

Chapter 4

Diversity Combining in Rayleigh Fading

Rayleigh distribution had been used for more than half a century [100] for modelling electromagnetic signal propagation through multipath wireless environments. Interestingly, far from being outdated, it remains the most acceptable model till now. There are mainly three reasons behind this: first, in urban and suburban land mobile communication hardly any direct LOS path exists between the transmitter (Tx) and the receiver (Rx). The mobile antenna receives a large number of reflected, diffracted and scattered waves from buildings, foliage, and rough terrain. In such an environment Rayleigh distribution closely matches with the channel attenuation profile as available from several field measurements. Second, Rayleigh is a good fit for not only the cellular communication channel, but also several other non-LOS (nLOS) propagations e.g., troposcatter and ionospheric propagation, maritime ship-to-ship communication, and underwater acoustic communication. Third, quite contrary to the complex physical basis of Rayleigh model, the PDF and associated formulas involved in the model are remarkably simple. Thus the end results of many performance assessment problems are mostly in closed-form for Rayleigh channels. In particular, compared to other fading models, the transmission error probability calculations are much simpler in this case. All these facts accelerated the rapid growth of research work based on Rayleigh models (Ref. [73] contains a detailed list of them).

In this chapter, we derived improvements of the error rates for various digital modulation systems with four major diversity combining techniques over single and double Rayleigh fading channels. The basic difference between these two fading models are outlined in Section 4.1. Next, in Section 4.2, a unified approach for calculating BEP of binary modulations with MRC, EGC, SC, and SWC (SSC as well as SEC) is presented. Calculations for the MPSK and ncMFSK for Rayleigh channels are trivial, and had been addressed by many researchers in the past. Thus, we restrict our discussion to other important M-ary modulation schemes like MDPSK, MFSK, and MQAM. Section 4.3 describes the BEP calculation for $\pi/4$-DQPSK, the most widely used modulation in mobile communication that belongs to the MDPSK family, when used with MRC, SC, EGC, or SSC combining methods. In Section 4.4, improved approximations for coherent MFSK with MRC and SC receivers are presented. Section 4.5, on the other hand, is concerned with SEP calculations for MQAM with SC and EGC combining. A simplified error expression for the EGC case is also mentioned, and the corresponding percentage of error, with respect to the exact solution, is graphically displayed. Finally we address the double Rayleigh fading channel in Section 4.6, where the BEP of binary modulations, with and without switched combining, is discussed.

4.1 Single and Double Rayleigh Fading Channel

Traditionally Rayleigh and other statistical distributions (Rice, Nakagami, Hoyt etc.) were introduced to model *fixed-to-mobile* (F2M) cellular radio channels, where the base-station is stationary, elevated, and relatively free from local scattering. However in *mobile-to-mobile* (M2M) communication, the conventional statistical models are not strictly applicable due to the mobility of transmitter and receiver as well as obstructions around their low elevation antennas. One simple example of such an M2M wireless scenario may be some emergency, where military/ rescue squad vehicles form a vehicular ad-hoc network (VANET) and each vehicle needs to constantly communicate with all other moving vehicles. M2M situations can also be observed in propagation via diffracting wedges (such as street corners) in urban micro cells, in ad-hoc mobile networks, dedicated short-range communication systems for intelligent highways, relay-based cellular networks and so on.

The received envelope in M2M wireless channels is normally modelled using a *double* or *cascaded* Rayleigh distribution under nLOS conditions [101, 102]. Existence of this double Rayleigh scattering phenomenon has been established through various channel measurement experiments [103, 104, 105, 106, 107, 108] and validated by some simulation studies [109, 110]. The baseband equivalent channel transfer function in a double Rayleigh fading model is given by a product of two independent zero-mean complex Gaussian RVs [111]. As shown in Fig. 4.1, the double fading model is based on two-ring path geometry [112, 113] where each ring consists of independent scatterers around the mobile terminals. All the scattered waves travel between these two rings through a narrow pipe like channel, popularly known as *keyhole* or *pinhole*. When a system employs multiple antennas to achieve diversity, keyhole effect induces uncorrelated spatial fading at both ends of the channel, and is likely to improve the system performance. However, the channel transfer matrix becomes rank-deficient (have a single or reduced degree of freedom) in this mode of

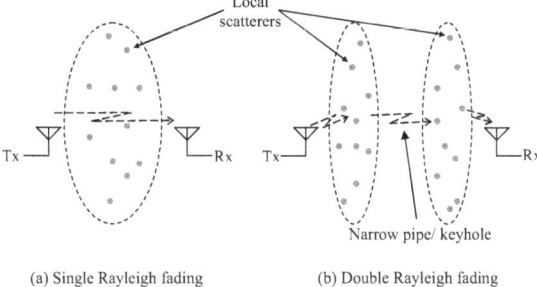

(a) Single Rayleigh fading   (b) Double Rayleigh fading

Figure 4.1: Physical realization of single and double Rayleigh fading. In the absence of direct LOS path, (a) when the scatterers are uniformly distributed between Tx and Rx, signal attenuation follows single Rayleigh distribution, and (b) when scatterers are distributed around Tx and Rx and the scatter rings are small compared to the distance between Tx and Rx, signal attenuation follows double (or cascaded) Rayleigh fading.

propagation. This rank deficiency in keyhole channels reduces effective channel capacity and severely degrades the link quality in MIMO systems [103, 104].

4.2 Performance of Binary Modulations

In this section, using the unified expression as devised by Wojner [7], an analytical framework is presented to determine error performance of all binary modulation schemes employing coherent or non-coherent detection and operating over slow flat IID Rayleigh fading channel with different diversity combining (SC, EGC, MRC, and SWC). A class of generalized expressions for BEP has been obtained following PDF or CDF approach. This approach simplifies the calculations since a single formula is used to derive BEP for different modulation schemes and thus dispensing with the need for BEP calculation for individual modulation schemes separately.

4.2.1 Unified Approach for BEP Analysis

The most commonly used technique for this task is the PDF based method, where BEP can be obtained as

$$P_{e,b} = \int_0^\infty P_{e,b}(\gamma_b) f_\gamma(\gamma_b) d\gamma_b \tag{4.1}$$

where, γ_b is the instantaneous SNR per bit at the combiner output, $f_\gamma(\cdot)$ denotes the corresponding PDF in a specified fading environment, and $P_{e,b}(\gamma_b)$ is the conditional BEP under the said fading scenario. Wojner [7] presented a general expression for conditional error probability of binary signalling schemes in terms of complementary incomplete gamma function

$$P_{e,b}(\gamma_b) = \frac{\Gamma(\beta, \alpha\gamma_b)}{2\Gamma(\beta)} \qquad (4.2)$$

where α denotes type of constellation (1/2 for orthogonal and 1 for antipodal) and β accounts for the type of detection (1/2 for coherent and 1 for non-coherent).

In some cases, evaluation of BEP is more convenient with CDF of the combiner output SNR instead of the PDF. Fortunately there exists an alternate representation of (4.1) in terms of the derivative $P'_{e,b}(\gamma_b)$ and the CDF $F_\gamma(\gamma_b)$ of SNR γ_b

$$P_{e,b} = -\int_0^\infty P'_{e,b}(\gamma_b)F_\gamma(\gamma_b)d\gamma_b \qquad (4.3)$$

which can be derived by integrating the integral in (4.1) by parts and using the limiting values $F_{\gamma_b}(0) = P_{e,b}(\infty) = 0$. Using the relation [9, (6.5.25)], $\partial\Gamma(a,z)/\partial z = -z^{a-1}\exp(-z)$, one can identify that the first derivative $P'_{e,b}(\gamma_b)$ is composed of a product of finite power and exponential function of γ_b

$$P'_{e,b}(\gamma_b) = -\frac{\alpha^\beta}{2\Gamma(\beta)}\gamma_b^{\beta-1}\exp(-\alpha\gamma_b) \qquad (4.4)$$

which, when inserted in (4.3), greatly simplifies the integration process. Substituting the CDF from (3.40) and utilizing the relation $\int_0^\infty z^{n-1}\exp(-\beta z)dz = \Gamma(n)/\beta^n$, we have

$$
\begin{aligned}
P_{e,b} &= \frac{\alpha^\beta}{2\Gamma(\beta)}\left[\int_0^\infty \gamma_b^{\beta-1}\exp(-\alpha\gamma_b)d\gamma_b \right. \\
&\qquad \left. -\int_0^\infty \gamma_b^{\beta-1}\exp\left(-\gamma_b\frac{1+\alpha\bar{\gamma}_b}{\bar{\gamma}_b}\right)d\gamma_b\right] \\
&= \frac{1}{2}\left[1 - \left(\frac{\alpha\bar{\gamma}_b}{1+\alpha\bar{\gamma}_b}\right)^\beta\right]
\end{aligned}
\qquad (4.5)
$$

giving the unified average BEP for all four kind of binary modulations over Rayleigh fading channel. Alternatively, by averaging (4.2) over the Rayleigh PDF while making use of a definite integration formula [9, (6.5.36)]

$$\int_0^\infty \exp(-ax)\Gamma(b, tx)dx = \frac{\Gamma(b)}{a}\left[1 - \left(\frac{t}{a+t}\right)^b\right] \qquad (4.6)$$

the BEP expression given in (4.5) may be obtained.

4.2.2 BEP for MRC

For L-branch MRC diversity over IID Rayleigh channels, the PDF of γ_b is chi-square distributed with $2L$ degrees of freedom as mentioned in (3.85). Now substituting (3.85) and (4.2) into (4.1) and using [8, (6.455.1)], we obtain BEP expression in the form

$$
\begin{aligned}
P_{e,b} &= \frac{1}{2L}\frac{\Gamma(L+\beta)}{\Gamma(L)\Gamma(\beta)}\frac{(\alpha\bar{\gamma}_b)^\beta}{(1+\alpha\bar{\gamma}_b)^{L+\beta}} \\
&\quad \times {}_2F_1\left(1, L+\beta; L+1; \frac{1}{1+\alpha\bar{\gamma}_b}\right)
\end{aligned}
\tag{4.7}
$$

where, ${}_2F_1(\cdot,\cdot;\cdot;\cdot)$ denotes the Gauss hypergeometric function [8, (9.100)]. For a quick check, one may put $L = 1$ (for the no diversity case) in (4.7) and thereafter apply [8, (9.121.5)] to get back (4.5).

4.2.3 BEP for EGC

In case of EGC, we prefer the CDF based approach defined in (4.3) as the CDF of combiner output SNR involves lesser terms compared to the corresponding PDF expression. The CDF of γ_b for dual diversity EGC over IID Rayleigh channels is given in (3.93). Now putting (4.4) and (3.93) into (4.3) and after some straightforward calculation we obtain

$$
\begin{aligned}
P_{e,b} &= \frac{\alpha^\beta}{2\Gamma(\beta)}\left\{\frac{\Gamma(\beta)}{\alpha^\beta} - \left(\frac{\bar{\gamma}_b}{2+\alpha\bar{\gamma}_b}\right)^\beta \Gamma(\beta)\right. \\
&\quad \left. -\sqrt{\frac{\pi}{\bar{\gamma}_b}}\left(\frac{\bar{\gamma}_b}{1+\alpha\bar{\gamma}_b}\right)^{\beta+1/2}\Gamma(\beta+\frac{1}{2}) + 2\sqrt{\frac{\pi}{\bar{\gamma}_b}}\mathcal{I}_1\right\}
\end{aligned}
\tag{4.8}
$$

where

$$
\mathcal{I}_1 = \frac{1}{\sqrt{2\pi}}\int_0^\infty \gamma_b^{\beta-1/2}\exp\left[-\frac{\gamma_b(1+\alpha\gamma_b)}{\bar{\gamma}_b}\right]Q\left(\sqrt{\frac{2\gamma_b}{\bar{\gamma}_b}}\right)d\gamma_b
\tag{4.9}
$$

From the definition of $Q(z) = (1/\sqrt{2\pi})\int_z^\infty \exp(u^2/2)du$, the integral can be rewritten in the form

$$
\begin{aligned}
\mathcal{I}_1 &= \frac{1}{\sqrt{2\pi}}\int_0^\infty \gamma_b^{\beta-1/2}\exp\left[-\frac{\gamma_b(1+\alpha\gamma_b)}{\bar{\gamma}_b}\right] \\
&\quad \times \left\{\int_{\sqrt{2\gamma_b/\bar{\gamma}_b}}^\infty \exp(-z^2)dz\right\}d\gamma_b
\end{aligned}
\tag{4.10}
$$

Changing the order of integral in (4.10) and using (6.455.1) [8], the expression of the integral I_1 becomes

$$
I_1 = \frac{1}{2}\left[\frac{\Gamma\left(\beta+\frac{1}{2}\right)}{v^{\beta+\frac{1}{2}}} - \frac{2\Gamma(\beta+1)}{\sqrt{\pi}\bar{\gamma}_b}\left(\frac{\bar{\gamma}_b}{1+v\bar{\gamma}_b}\right)^{\beta+1} \right.
$$
$$
\left. \times {}_2F_1\left(1,\beta+1;\frac{3}{2};\frac{1}{1+v\bar{\gamma}_b}\right)\right]
$$
(4.11)

where $v = \alpha + 1/\bar{\gamma}_b$. Substituting (4.11) in (4.8) and inserting the value of v, generalized expression for BEP with two-branch EGC in closed-form becomes

$$
P_{e,b} = \frac{1}{2}\left[1 - \left(\frac{\alpha\bar{\gamma}_b}{2+\alpha\bar{\gamma}_b}\right)^{\beta}\left\{1 + \frac{2\beta}{2+\alpha\bar{\gamma}_b}\right.\right.
$$
$$
\left.\left. \times {}_2F_1\left(1,\beta+1;\frac{3}{2};\frac{1}{2+\alpha\bar{\gamma}_b}\right)\right\}\right]
$$
(4.12)

Putting the parameter values for BPSK ($\alpha = 1, \beta = 1/2$) and using [8, (9.121.1)] it is easy to show after some algebra that (4.12) reduces to (3.96).

4.2.4 BEP for SC

Similar to the EGC case, the BEP evaluation process for SC is also simpler with CDF based approach. However, BEP in this case is derived for the general L-branch combining and not restricted to only dual diversity case. For L-branch SC diversity over IID Rayleigh channels, the CDF of γ_b is given by (3.75). Utilizing the relation

$$
[1 - \exp(-x)]^L = \sum_{k=0}^{L}\binom{L}{k}(-1)^k\exp(-kx)
$$
(4.13)

the CDF can be written in a more suitable form

$$
F_{\gamma_{SC}}(\gamma_b) = \sum_{k=0}^{L}\binom{L}{k}(-1)^k\exp\left(-\frac{k\gamma_b}{\bar{\gamma}_b}\right)
$$
(4.14)

Now inserting (4.4) and expression for the CDF given in (4.14) into (4.3) we obtain

$$
P_{e,b} = \frac{\alpha^{\beta}}{2\Gamma(\beta)}\sum_{k=0}^{L}\binom{L}{k}(-1)^k\int_0^{\infty}\gamma_b^{\beta-1}\exp\left[-\frac{\gamma(k+\alpha\bar{\gamma}_b)}{\bar{\gamma}_b}\right]d\gamma_b
$$
(4.15)

which, when evaluated reduces to the following form

$$
P_{e,b} = \frac{1}{2}\sum_{k=0}^{L}\binom{L}{k}(-1)^k\left(\frac{\alpha\bar{\gamma}_b}{k+\alpha\bar{\gamma}_b}\right)^{\beta}
$$
(4.16)

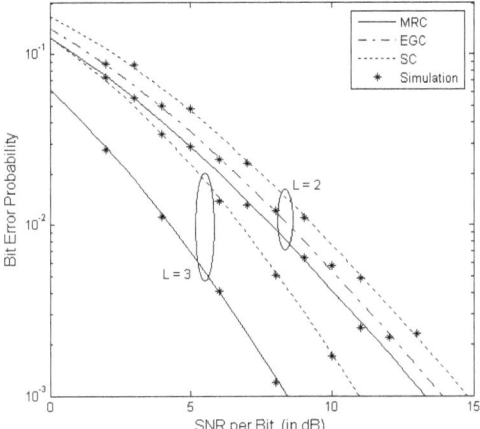

Figure 4.2: BEP of DPSK with different combining (SC, EGC, and MRC) in Rayleigh fading.

For dual-branch combining ($L = 2$), BEP performance of MRC, EGC, SC for DPSK modulation are depicted in Fig. 4.2. For SC and MRC, the results have also been shown for diversity orders $L = 3$. We notice that the results are similar to those obtained in earlier works [1, 2]. The results are, however, obtained in a much simpler form without considering each and every modulation scheme separately, and thereby avoiding unnecessary calculations. Further, the analytical results match perfectly with the simulated values. The details of simulation may be found in Chapter 7.6.5.

4.2.5 BEP for SSC

For dual-branch SSC over IID Rayleigh channels, the PDF of γ_b is given in (3.101). Now substituting (3.101) and (4.2) into (4.1) we obtain

$$P_{e,b} = \frac{1}{2\bar{\gamma}_b \Gamma(\beta)} \left[\left\{ 1 - \exp\left(-\frac{\gamma_T}{\bar{\gamma}_b} \right) \right\} \mathcal{I}_2 + \mathcal{I}_3 \right] \tag{4.17}$$

where

$$\mathcal{I}_2 = \int_0^\infty \Gamma(\beta, \alpha\gamma_b) \exp\left(-\frac{\gamma_b}{\bar{\gamma}_b} \right) \tag{4.18}$$

and

$$\mathcal{I}_3 = \int_{\gamma_T}^\infty \Gamma(\beta, \alpha\gamma_b) \exp\left(-\frac{\gamma_b}{\bar{\gamma}_b} \right) \tag{4.19}$$

$\mathcal{I}_2$ can be solved readily using [9, (6.5.36)] giving

$$\mathcal{I}_2 = \bar{\gamma}_b \Gamma(\beta) \left[1 - \left(\frac{\alpha \bar{\gamma}_b}{1 + \alpha \bar{\gamma}_b} \right)^{\beta} \right] \tag{4.20}$$

On the other hand, $\mathcal{I}_3$ solved through the method of integration by parts gives

$$\mathcal{I}_3 = \bar{\gamma}_b \left[\exp \left(-\frac{\gamma_T}{\bar{\gamma}_b} \right) \Gamma(\beta, \alpha \gamma_T) - \alpha^{\beta} \mathcal{I}_4 \right] \tag{4.21}$$

where

$$\mathcal{I}_4 = \left(\frac{\bar{\gamma}_b}{1 + \alpha \bar{\gamma}_b} \right)^{\beta} \Gamma \left[\beta, \frac{\gamma_T (1 + \alpha \bar{\gamma}_b)}{\bar{\gamma}_b} \right] \tag{4.22}$$

Finally the closed form expression for BEP is obtained as follows

$$
\begin{aligned}
P_{e,b} = \ & \frac{1}{2} \left[1 - \exp(-\hat{\gamma}_b) \left\{ 1 - \frac{\Gamma(\beta, \alpha \gamma_T)}{\Gamma(\beta)} \right\} \right. \\
& - \left(\frac{\alpha \bar{\gamma}_b}{1 + \alpha \bar{\gamma}_b} \right)^{\beta} \{ 1 - \exp(-\hat{\gamma}_b) \\
& \left. + \frac{\Gamma(\beta, \hat{\gamma}_b (1 + \alpha \bar{\gamma}_b))}{\Gamma(\beta)} \right\} \right]
\end{aligned}
\tag{4.23}
$$

where $\hat{\gamma}_b = \gamma_T / \bar{\gamma}_b$ denotes the normalized threshold.

Noting that $\Gamma(z, 0) = \Gamma(z)$ [8, (8.350.3)], one can identify that the SSC performance with $\gamma_T = 0$ becomes identical with the no diversity case in (4.5), for no switching occurring at all. Transmitting data at a fixed SNR i.e., keeping the value of average SNR $\bar{\gamma}_b$ constant, as the switching threshold increases from 0, the probability of switching

$$\Pr[\gamma_b < \gamma_T] = \int_0^{\gamma_T} f_{\gamma}(\gamma_b) d\gamma_b \tag{4.24}$$

also increases causing the BER to decrease until an optimum threshold SNR value γ_T^* is reached. Naturally, the optimum threshold γ_T^* ensures minimum average error probability and can be obtained from the following identity [114, (7)]

$$\left. \frac{\partial P_{e,b}}{\partial \gamma_T} \right|_{\gamma_T = \gamma_T^*} = 0 \tag{4.25}$$

In the range $\gamma_T > \gamma_T^*$ the BER starts increasing once again. Finally, for $\gamma_T \to \infty$, there will be constant switching because no branch can satisfy threshold requirement. In this case the performance will be equivalent to the performance of diversity branch selected at random, which resembles the behaviour of single branch diversity. Mathematically, (4.23) boils down to (4.5) once again as $\Gamma(z, \infty) = 0$ [8, (8.350.4)]. A quick look at (3.101) reveals that in both the cases i.e., for $\gamma_T = 0$ and $\gamma_T \rightarrow \infty$, (3.101) reduces to simple Rayleigh PDF.

Despite the fact that closed-form expressions for γ_T^* with specific signalling constellation can be easily obtained, a unified expression, including all the modulation schemes under consideration, cannot be extracted. Differentiating (4.23) with respect to γ_T, equating the differential to zero, and noting that $\Gamma(1, z) = \exp(-z)$ [8, (8.352.7)] and $\Gamma(1/2, z^2) = \sqrt{\pi}\,\mathrm{erfc}(z)$ [9, (6.5.17)], we get a set of two different equations. For non-coherent detection γ_T^* can be written directly as an analytic function of $\bar{\gamma}_b$

$$\gamma_T^* = \frac{1}{\alpha} \ln(1 + \alpha\bar{\gamma}_b) \tag{4.26}$$

while for coherent detection γ_T^* is

$$\gamma_T^* = \frac{1}{\alpha}\left[\mathrm{erfcinv}\left(1 - \sqrt{\frac{\alpha\bar{\gamma}_b}{1 + \alpha\bar{\gamma}_b}}\right)\right]^2 \tag{4.27}$$

where $\mathrm{erfcinv}(\cdot)$ denotes inverse complementary error function.

4.2.6 BEP for SEC

The PDF $f_{\gamma_{SEC}}(\gamma)$ for L branch SEC under Rayleigh fading is given as [115]

$$f_{\gamma_{SEC}}(\gamma) = \begin{cases} \dfrac{1}{\bar{\gamma}}\exp\left(-\dfrac{\gamma}{\bar{\gamma}}\right)\left[1 - \exp\left(-\dfrac{\gamma_T}{\bar{\gamma}}\right)\right]^{L-1} & ; \gamma < \gamma_T \\[2mm] \dfrac{1}{\bar{\gamma}}\exp\left(-\dfrac{\gamma}{\bar{\gamma}}\right)\displaystyle\sum_{k=0}^{L-1}\left[1 - \exp\left(-\dfrac{\gamma_T}{\bar{\gamma}}\right)\right]^{k} & ; \gamma \geq \gamma_T \end{cases} \tag{4.28}$$

which is same as (3.101) for $L = 2$. This observation mathematically proves that L branch SSC is just a special case of SEC for $L = 2$. Inserting $P_{e,b}(\gamma_b)$ given by (4.2) and $f_{\gamma_{SEC}}(\gamma)$ from (4.28) in (4.1) gives

$$P_{e,b} = \frac{\mathcal{I}_1}{2\bar{\gamma}_b\Gamma(\beta)}\left[1 - \exp(-\hat{\gamma})\right]^{L-1} + \frac{\mathcal{I}_2}{2\bar{\gamma}_b\Gamma(\beta)}\sum_{k=0}^{L-2}\left[1 - \exp(-\hat{\gamma})\right]^{k} \tag{4.29}$$

where $\mathcal{I}_1 = \int_0^\infty \Psi(\gamma_b)d\gamma_b$, $\mathcal{I}_2 = \int_{\gamma_T}^\infty \Psi(\gamma_b)d\gamma_b$, $\hat{\gamma} = \gamma_T/\bar{\gamma}_b$, and $\Psi(\gamma_b) = \Gamma(\beta, \alpha\gamma_b)\exp(-\gamma_b/\bar{\gamma}_b)$. Using (4.6), the first integral can be readily evaluated as

$$\mathcal{I}_1 = \Gamma(\beta)\bar{\gamma}_b\left[1 - \left(\frac{\alpha\bar{\gamma}_b}{1 + \alpha\bar{\gamma}_b}\right)^{\beta}\right] \tag{4.30}$$

However, calculation of the second integral is not straightforward like the first as the lower limit of the integration, γ_T, is having a finite non-zero value. Solving the same through the method of integration by parts, the result can be expressed with yet another complementary incomplete gamma function

$$\mathcal{I}_2 = \bar{\gamma}_b \Gamma(\beta, \alpha \gamma_T) \exp(-\hat{\gamma}) - \bar{\gamma}_b \left(\frac{\alpha \bar{\gamma}_b}{1 + \alpha \bar{\gamma}_b} \right)^\beta \Gamma[\beta, \hat{\gamma}(1 + \alpha \bar{\gamma}_b)] \tag{4.31}$$

Finally, accumulating the results from (4.29)-(4.31), the overall unified BER expression for binary modulations with SEC can be written as

$$\begin{aligned}
P_{e,b} &= \frac{1}{2} [1 - \exp(-\hat{\gamma}_b)]^{L-1} \left[1 - \left(\frac{\alpha \bar{\gamma}_b}{1 + \alpha \bar{\gamma}_b} \right)^\beta \right] \\
&\quad + \frac{1}{2\Gamma(\beta)} \sum_{k=0}^{L-2} [1 - \exp(-\hat{\gamma}_b)]^k \left\{ \Gamma(\beta, \alpha \gamma_T) \exp(-\hat{\gamma}_b) \right. \\
&\quad \left. - \left(\frac{\alpha \bar{\gamma}_b}{1 + \alpha \bar{\gamma}_b} \right)^\beta \Gamma[\beta, \hat{\gamma}_b(1 + \alpha \bar{\gamma}_b)] \right\}
\end{aligned} \tag{4.32}$$

involving only elementary functions. Note that, for $L - 2$, (4.32) reduces to (4.23) as expected.

$\bar{\gamma}_b$	Optimum common switching threshold γ_T^* (dB)					
(dB)	BPSK ($\alpha = 1$, $\beta = \frac{1}{2}$)			DPSK($\alpha = 1$, $\beta = 1$)		
	$L=8$	$L=4$	SSC	$L=8$	$L=4$	SSC
0	0.84	-0.71	-2.57	1.33	-0.03	-1.59
3	3.08	1.43	-0.54	3.53	2.08	0.40
6	5.05	3.28	1.16	5.47	3.88	2.06
9	6.77	4.85	2.56	7.15	5.41	3.41
12	8.24	6.19	3.71	8.61	6.70	4.51
15	9.58	7.32	4.67	9.90	7.81	5.42
18	10.73	8.21	5.49	11.03	8.71	6.19
21	11.73	9.20	6.18	11.99	9.60	6.85
24	12.61	9.96	6.80	12.86	10.34	7.43
27	13.39	10.63	7.34	13.61	10.99	7.94

Table 4.1: Optimum switching threshold (γ_T^*) for BPSK and DPSK with SEC in Rayleigh fading.

For SEC the optimal threshold γ_T^* is an increasing function of both $\bar{\gamma}_b$ and L. Indicative values of the optimum switching threshold γ_T^* for minimum BER are computed for BPSK and DPSK modulations through numerical minimization using $fminunc()$ function available in the well-known mathematical software package MATLAB and are listed in Table 4.1.

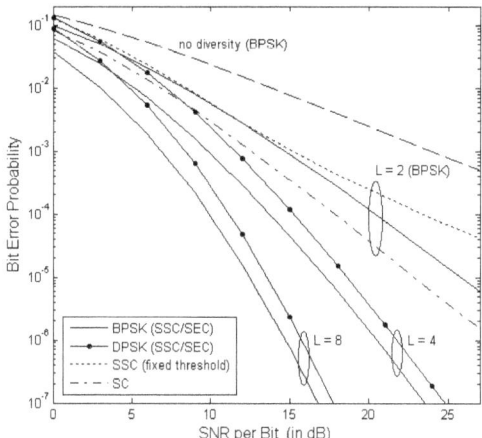

Figure 4.3: BEP of BPSK and DPSK with SSC/ SEC in Rayleigh fading.

The BER performances of BPSK and DPSK with SSC and SEC (operating with common optimum switching threshold) in Rayleigh fading channel are presented in Fig. 4.3. The BER values are shown for different orders of diversity ($L = 2$, 4, and 8). When L is greater than two ($L > 2$), error performance of a system with SEC diversity improves while for the systems using SSC diversity, the performance remains same. For comparison purpose, BER of BPSK for the no diversity case ($L = 1$), with dual diversity SC, and with SSC (or SEC with $L = 2$) when the threshold value is fixed ($\gamma_T = 3dB$) have also been shown. It is easily recognized that for SSC/SEC, systems with optimum threshold outperform those with fixed threshold. However, SC error rates are still well below any kind of SWC. Simulation results for all these cases are shown later, alongwith the double Rayleigh results in Section 4.6.2.

It may be noted that part of some derivations presented in this section are somewhat similar to those obtained in earlier work [73]. Specifically, for the BPSK case ($\alpha = 1, \beta = 1/2$), (4.23) and (4.32) reduce to (9.306) and (9.343) of [73] and (4.27) becomes (9.313) [73] for $\alpha = 1$. In this book the results are, however, obtained in a unified form without considering each and every modulation scheme separately, and thereby avoiding repeated calculations. For SEC, the optimum threshold values have been evaluated as given in Table 4.1 although nothing about optimum threshold with SEC is discussed in [73].

4.3 Performance of $\pi/4$-DQPSK

Using the infinite series expressions of Marcum's Q function [36], the conditional BEP for $\pi/4$-DQPSK can be written from (2.94) as

$$
\begin{aligned}
P_{e,b}(\gamma_b) &= \exp(-2\gamma_b) \sum_{j=0}^{\infty} \left(\frac{1}{\sqrt{2}+1} \right)^j I_j \left(\sqrt{2}\gamma_b \right) \\
&\quad - \frac{1}{2} \exp(-2\gamma_b) I_0 \left(\sqrt{2}\gamma_b \right)
\end{aligned}
\tag{4.33}
$$

and subsequently the expression for average BEP can be derived by averaging this conditional BEP over the PDF of γ_b

$$
P_{e,b} = \int_0^{\infty} P_{e,b}(\gamma_b) f_{\gamma}(\gamma_b) d\gamma_b = \mathcal{I}_1 - \mathcal{I}_2
\tag{4.34}
$$

where

$$
\mathcal{I}_1 = \frac{1}{\bar{\gamma}_b} \sum_{j=0}^{\infty} \left(\frac{1}{\sqrt{2}+1} \right)^j \int_0^{\infty} \exp\left[-\gamma_b \left(2 + \frac{1}{\bar{\gamma}_b} \right) \right] I_j \left(\sqrt{2}\gamma_b \right) d\gamma_b
\tag{4.35}
$$

and

$$
\mathcal{I}_2 = \frac{1}{2\bar{\gamma}_b} \int_0^{\infty} \exp\left[-\gamma_b \left(2 + \frac{1}{\bar{\gamma}_b} \right) \right] I_0 \left(\sqrt{2}\gamma_b \right) d\gamma_b
\tag{4.36}
$$

Using [8, (6.611.4)] the first integral $\mathcal{I}_1$ can be evaluated as

$$
\mathcal{I}_1 = \frac{1}{\sqrt{2\bar{\gamma}_b^2 + 4\bar{\gamma}_b + 1}} \sum_{j=0}^{\infty} \left[\frac{(2\bar{\gamma}_b + 1) - \sqrt{2\bar{\gamma}_b^2 + 4\bar{\gamma}_b + 1}}{\sqrt{2}(\sqrt{2}+1)\bar{\gamma}_b} \right]^j
\tag{4.37}
$$

which may be further simplified, noting that $\sum_{j=0}^{\infty} z^j = 1/(1-z); |z| \leq 1$, as

$$
\mathcal{I}_1 = \frac{1}{2} \left(1 - \frac{\sqrt{2}\bar{\gamma}_b - 1}{\sqrt{2\bar{\gamma}_b^2 + 4\bar{\gamma}_b + 1}} \right)
\tag{4.38}
$$

For the second integral we again apply [8, (6.611.4)] to obtain

$$
\mathcal{I}_2 = \frac{1}{2} \frac{1}{\sqrt{2\bar{\gamma}_b^2 + 4\bar{\gamma}_b + 1}}
\tag{4.39}
$$

Finally inserting (4.38) and (4.39) in (4.34), the total average BEP is

$$
P_{e,b} = \frac{1}{2} \left(1 - \sqrt{\frac{2\bar{\gamma}_b^2}{1 + 4\bar{\gamma}_b + 2\bar{\gamma}_b^2}} \right)
\tag{4.40}
$$

which agrees with (8) [38] and with (8.208) [73]. Equivalent expressions were also provided by Tjhung et al. [116] and Tanda [117]

$$
\begin{aligned}
P_{e,b} &= \frac{1}{2\sqrt{1 + 4\bar{\gamma}_b + 2\bar{\gamma}_b^2}} \\
&\times \left[\frac{\sqrt{2}\bar{\gamma}_b + \left(\sqrt{2} - 1\right)\left(1 + 2\bar{\gamma}_b - \sqrt{1 + 4\bar{\gamma}_b + 2\bar{\gamma}_b^2}\right)}{\sqrt{2}\bar{\gamma}_b - \left(\sqrt{2} - 1\right)\left(1 + 2\bar{\gamma}_b - \sqrt{1 + 4\bar{\gamma}_b + 2\bar{\gamma}_b^2}\right)} \right]
\end{aligned}
\tag{4.41}
$$

It may also be noted that (4.40) can be obtained directly by averaging the conditional BEP in the form as given in (2.93)

$$
\begin{aligned}
P_{e,b}(\gamma_b) &= \frac{1}{2}\left[1 - Q\left(\sqrt{\gamma_b(2 + \sqrt{2})}, \sqrt{\gamma_b(2 - \sqrt{2})}\right)\right. \\
&\left. +Q\left(\sqrt{\gamma_b(2 - \sqrt{2})}, \sqrt{\gamma_b(2 + \sqrt{2})}\right)\right]
\end{aligned}
\tag{4.42}
$$

over the Rayleigh PDF in (3.39) using (A.5).

4.3.1 BEP for MRC

For L-branch MRC diversity over IID Rayleigh channels, the PDF of γ_b is chi-square distributed with $2L$ degrees of freedom and given by (3.85)

$$
f_{\gamma_{MRC}}(\gamma_b) = \frac{1}{(L-1)!} \frac{\gamma_b^{L-1}}{\bar{\gamma}_b^L} \exp\left(-\frac{\gamma_b}{\bar{\gamma}_b}\right)
\tag{4.43}
$$

assuming each branch has equal average branch SNR $\bar{\gamma}_b = E\{\gamma_{b,k}\}$; $k = 1, 2, \cdots, L$. Now averaging (4.42) over (4.43) using (A.4) the average BEP of $\pi/4$-DQPSK with Lth order MRC diversity over Rayleigh fading channel becomes

$$
P_{e,b} = \frac{1}{2}\left[1 - \sum_{k=0}^{L-1} \frac{\sqrt{2}\bar{\gamma}_b}{(1 + 2\bar{\gamma}_b)^{k+1}} \, _2F_1\left(\frac{k+1}{2}, \frac{k+2}{2}; 1; \frac{2\bar{\gamma}_b^2}{(1 + 2\bar{\gamma}_b)^2}\right)\right]
\tag{4.44}
$$

or equivalently

$$
P_{e,b} = \frac{1}{2}\left[1 - \sum_{k=1}^{L} \frac{\sqrt{2}\bar{\gamma}_b}{(1 + 2\bar{\gamma}_b)^k} \, _2F_1\left(\frac{k}{2}, \frac{k+1}{2}; 1; \frac{2\bar{\gamma}_b^2}{(1 + 2\bar{\gamma}_b)^2}\right)\right]
\tag{4.45}
$$

4.3.2 BEP for SC

Utilizing the binomial series expansion of $(1 - x)^{L-1} = \sum_{k=0}^{L-1} \binom{L-1}{k}(-1)^k x^k$, the PDF for selection combiner output SNR in IID Rayleigh fading channel given in (3.76) can be written in a more suitable form as

$$f_{\gamma_{SC}}(\gamma_b) = \frac{L}{\bar{\gamma}_b} \sum_{k=0}^{L-1} \binom{L-1}{k} (-1)^k \exp\left[-\gamma_b \left(\frac{k+1}{\bar{\gamma}_b}\right)\right] \qquad (4.46)$$

Now averaging (4.42) over (4.46) using (A.5), the average BEP of $\pi/4$-DQPSK with Lth order SC diversity over Rayleigh fading channel becomes

$$P_{e,b} = \frac{1}{2}\left[1 - \sum_{k=0}^{L-1} \binom{L}{k+1}(-1)^k \left\{\frac{\sqrt{2}\bar{\gamma}_b}{\sqrt{(k+1)^2 + 4(k+1)\bar{\gamma}_b + 2\bar{\gamma}_b^2}}\right\}\right] \qquad (4.47)$$

or equivalently

$$P_{e,b} = \frac{1}{2}\left[1 + \sum_{k=1}^{L} \binom{L}{k}(-1)^k \sqrt{\frac{2\bar{\gamma}_b^2}{k^2 + 4k\bar{\gamma}_b + 2\bar{\gamma}_b^2}}\right] \qquad (4.48)$$

4.3.3 BEP for EGC

For dual diversity EGC over IID Rayleigh fading channel, the PDF of γ_b is given by (3.95)

$$\begin{aligned} f_{\gamma_{EGC}}(\gamma_b) &= \frac{1}{\bar{\gamma}_b}\exp\left(-\frac{2\gamma_b}{\bar{\gamma}_b}\right) - \frac{1}{2}\sqrt{\frac{\pi}{\bar{\gamma}_b}}\frac{1}{\sqrt{\gamma_b}}\exp\left(-\frac{\gamma_b}{\bar{\gamma}_b}\right) \\ &\quad + \frac{\sqrt{\pi}}{\bar{\gamma}_b^{3/2}}\sqrt{\gamma_b}\exp\left(-\frac{\gamma_b}{\bar{\gamma}_b}\right) \\ &\quad + \frac{1}{2}\sqrt{\frac{\pi}{\bar{\gamma}_b}}\frac{1}{\sqrt{\gamma_b}}\exp\left(-\frac{\gamma_b}{\bar{\gamma}_b}\right)\operatorname{erfc}\left(\sqrt{\frac{\gamma_b}{\bar{\gamma}_b}}\right) \\ &\quad - \frac{\sqrt{\pi}}{\bar{\gamma}_b^{3/2}}\sqrt{\gamma_b}\exp\left(-\frac{\gamma_b}{\bar{\gamma}_b}\right)\operatorname{erfc}\left(\sqrt{\frac{\gamma_b}{\bar{\gamma}_b}}\right) \end{aligned} \qquad (4.49)$$

Now using the conditional error probability given in the form of finite integral in (2.97), i.e.

$$P_{e,b}(\gamma_b) = \frac{1}{2\pi}\int_0^\pi \frac{\exp[-\gamma_b(2 - \sqrt{2}\cos\theta)]}{\sqrt{2} - \cos\theta}d\theta \qquad (4.50)$$

the average error probability can be found through the following integral

$$P_{e,b} = \frac{1}{2\pi}\int_0^\pi \frac{1}{\sqrt{2} - \cos\theta}[\mathcal{I}_1 - \mathcal{I}_2 + \mathcal{I}_3 + \mathcal{I}_4 - \mathcal{I}_5]d\theta \qquad (4.51)$$

where

$$\mathcal{I}_1 = \frac{1}{\bar{\gamma}_b} \int_0^\infty \exp\left[-\gamma_b \frac{2 + \bar{\gamma}_b(2 - \sqrt{2}\cos\theta)}{\bar{\gamma}_b}\right] d\gamma_b$$

$$= \frac{1}{2 + \bar{\gamma}_b(2 - \sqrt{2}\cos\theta)} \tag{4.52}$$

$$\mathcal{I}_2 = \frac{1}{2}\sqrt{\frac{\pi}{\bar{\gamma}_b}} \int_0^\infty \frac{1}{\sqrt{\gamma_b}} \exp\left[-\gamma_b \frac{1 + \bar{\gamma}_b(2 - \sqrt{2}\cos\theta)}{\bar{\gamma}_b}\right] d\gamma_b$$

$$= \frac{\pi}{2\sqrt{1 + \bar{\gamma}_b(2 - \sqrt{2}\cos\theta)}} \tag{4.53}$$

$$\mathcal{I}_3 = \frac{\sqrt{\pi}}{\bar{\gamma}_b^{3/2}} \int_0^\infty \sqrt{\gamma_b} \exp\left[-\gamma_b \frac{1 + \bar{\gamma}_b(2 - \sqrt{2}\cos\theta)}{\bar{\gamma}_b}\right] d\gamma_b$$

$$= \frac{\pi}{2}\left[\frac{1}{1 + \bar{\gamma}_b(2 - \sqrt{2}\cos\theta)}\right]^{3/2} \tag{4.54}$$

are evaluated using the relation $\int_0^\infty z^{n-1}\exp(-\beta z)dz = \Gamma(n)/\beta^n$. The next two terms are

$$\mathcal{I}_4 = \frac{1}{2}\sqrt{\frac{\pi}{\bar{\gamma}_b}} \int_0^\infty \frac{1}{\sqrt{\gamma_b}}\text{erfc}\left(\sqrt{\frac{\gamma_b}{\bar{\gamma}_b}}\right)$$

$$\times \exp\left[-\gamma_b \frac{1 + \bar{\gamma}_b(2 - \sqrt{2}\cos\theta)}{\bar{\gamma}_b}\right] d\gamma_b$$

$$= \frac{1}{\sqrt{1 + \bar{\gamma}_b(2 - \sqrt{2}\cos\theta)}} \tan^{-1}\left(\sqrt{1 + \bar{\gamma}_b(2 - \sqrt{2}\cos\theta)}\right) \tag{4.55}$$

and

$$\mathcal{I}_5 = \frac{\sqrt{\pi}}{\bar{\gamma}_b^{3/2}} \int_0^\infty \sqrt{\gamma_b}\text{erfc}\left(\sqrt{\frac{\gamma_b}{\bar{\gamma}_b}}\right)$$

$$\times \exp\left[-\gamma_b \frac{1 + \bar{\gamma}_b(2 - \sqrt{2}\cos\theta)}{\bar{\gamma}_b}\right] d\gamma_b$$

$$= \frac{1}{[1 + \bar{\gamma}_b(2 - \sqrt{2}\cos\theta)]^3} \tan^{-1}\left(\sqrt{1 + \bar{\gamma}_b(2 - \sqrt{2}\cos\theta)}\right)$$

$$- \frac{1}{[1 + \bar{\gamma}_b(2 - \sqrt{2}\cos\theta)][2 + \bar{\gamma}_b(2 - \sqrt{2}\cos\theta)]} \tag{4.56}$$

which are calculated utilizing [8, (6.285.1)] and [8, (6.292)] respectively. Finally inserting the expressions for $\mathcal{I}_1$ to $\mathcal{I}_5$ from (4.52)-(4.56) in (4.51) we have

$$
\begin{aligned}
P_{e,b} &= \frac{1}{2\pi} \int_0^\pi \frac{1}{(\sqrt{2} - \cos\theta)[1 + \bar{\gamma}_b(2 - \sqrt{2}\cos\theta)]} \\
&\quad \times \left[1 - \frac{\bar{\gamma}_b(2 - \sqrt{2}\cos\theta)}{\sqrt{1 + \bar{\gamma}_b(2 - \sqrt{2}\cos\theta)}} \right. \\
&\quad \left. \times \left\{ \frac{\pi}{2} - \tan^{-1}\left(\sqrt{1 + \bar{\gamma}_b(2 - \sqrt{2}\cos\theta)}\right) \right\} \right] d\theta
\end{aligned}
\tag{4.57}
$$

or equivalently

$$
\begin{aligned}
P_{e,b} &= \frac{1}{2\pi} \int_0^\pi \frac{1}{D(\sqrt{2} - \cos\theta)} \\
&\quad \times \left[1 - \frac{D-1}{\sqrt{D}} \left\{ \frac{\pi}{2} - \tan^{-1}(\sqrt{D}) \right\} \right] d\theta \\
&= \frac{\bar{\gamma}_b}{\sqrt{2}\pi} \int_0^\pi \frac{1}{D} \left[\frac{1}{D-1} - \frac{1}{\sqrt{D}} \tan^{-1}\left(\frac{1}{\sqrt{D}}\right) \right] d\theta
\end{aligned}
\tag{4.58}
$$

where $D = 1 + \bar{\gamma}_b(2 - \sqrt{2}\cos\theta)$.

4.3.4 BEP for SSC

For dual-branch SSC over IID Rayleigh channels, the PDF of γ_b is given by (3.101)

$$
f_{\gamma_{SSC}}(\gamma_b) = \begin{cases} \dfrac{1}{\bar{\gamma}_b} \exp\left(-\dfrac{\gamma_b}{\bar{\gamma}_b}\right) \left[1 - \exp\left(-\dfrac{\gamma_T}{\bar{\gamma}_b}\right) \right] & ; \gamma_b < \gamma_T \\[2ex] \dfrac{1}{\bar{\gamma}_b} \exp\left(-\dfrac{\gamma_b}{\bar{\gamma}_b}\right) \left[2 - \exp\left(-\dfrac{\gamma_T}{\bar{\gamma}_b}\right) \right] & ; \gamma_b \geq \gamma_T \end{cases}
\tag{4.59}
$$

where $\bar{\gamma}_b$ is the common average SNR per branch, i.e. $\bar{\gamma}_{b,1} = \bar{\gamma}_{b,2} = \bar{\gamma}_b$ and γ_T is the threshold SNR value. Now using the conditional error probability given in the form of finite integral in (4.50), the average error probability can be found as

$$
\begin{aligned}
P_{e,b} &= \frac{1}{2\pi\bar{\gamma}_b} \int_{\gamma_b=0}^{\gamma_T} \left[1 - \exp\left(-\frac{\gamma_T}{\bar{\gamma}_b}\right) \right] \exp\left(-\frac{\gamma_b}{\bar{\gamma}_b}\right) \\
&\quad \times \int_{\theta=0}^\pi \frac{\exp\left[-\gamma_b(2 - \sqrt{2}\cos\theta)\right]}{\sqrt{2} - \cos\theta} d\theta d\gamma_b \\
&\quad + \frac{1}{2\pi\bar{\gamma}_b} \int_{\gamma_b=\gamma_T}^\infty \left[2 - \exp\left(-\frac{\gamma_T}{\bar{\gamma}_b}\right) \right] \exp\left(-\frac{\gamma_b}{\bar{\gamma}_b}\right) \\
&\quad \times \int_{\theta=0}^\pi \frac{\exp\left[-\gamma_b(2 - \sqrt{2}\cos\theta)\right]}{\sqrt{2} - \cos\theta} d\theta d\gamma_b
\end{aligned}
\tag{4.60}
$$

Interchanging the order of integration, we have

$$
\begin{aligned}
P_{e,b} &= \frac{1 - \exp\left(-\frac{\gamma_T}{\bar{\gamma}_b}\right)}{2\pi\bar{\gamma}_b} \int_0^\pi \frac{1}{\sqrt{2} - \cos\theta} \\
&\quad \times \int_0^{\gamma_T} \exp\left[-\gamma_b \frac{1 + \bar{\gamma}_b(2 - \sqrt{2}\cos\theta)}{\bar{\gamma}_b}\right] d\gamma_b d\theta \\
&\quad + \frac{2 - \exp\left(-\frac{\gamma_T}{\bar{\gamma}_b}\right)}{2\pi\bar{\gamma}_b} \int_0^\pi \frac{1}{\sqrt{2} - \cos\theta} \\
&\quad \times \int_{\gamma_T}^\infty \exp\left[-\gamma_b \frac{1 + \bar{\gamma}_b(2 - \sqrt{2}\cos\theta)}{\bar{\gamma}_b}\right] d\gamma_b d\theta
\end{aligned}
\tag{4.61}
$$

Evaluation of the inner integrals and subsequent algebraic manipulation results in

$$
\begin{aligned}
P_{e,b} &= \frac{1}{2\pi} \int_0^\pi \frac{1}{\sqrt{2} - \cos\theta} \\
&\quad \times \frac{1 - \exp\left(-\frac{\gamma_T}{\bar{\gamma}_b}\right)\left\{1 - \exp\left[-\gamma_T\left(2 - \sqrt{2}\cos\theta\right)\right]\right\}}{1 + \bar{\gamma}_b(2 - \sqrt{2}\cos\theta)} d\theta
\end{aligned}
\tag{4.62}
$$

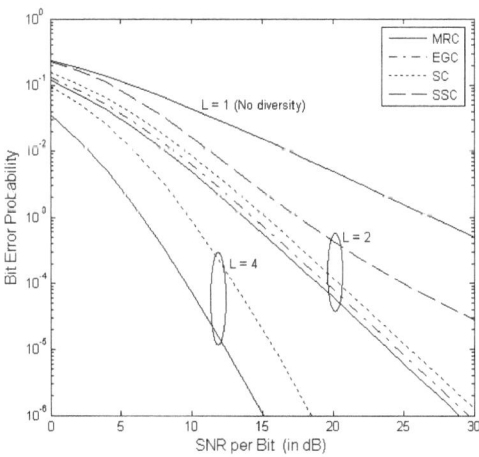

Figure 4.4: BEP of $\pi/4$-DQPSK in Rayleigh fading with MRC, EGC, SC and SSC.

In Fig. 4.4 the BEP expressions derived in previous subsections are plotted for two different diversity orders, $L = 2$ and 4. Simulation results are not shown in the figure as they would have unnecessarily cluttered the diagram. In the following chapters, simulation results for $\pi/4$-DQPSK modulation with different diversity combining schemes are displayed, and as Rayleigh fading is a special case for all the subsequent fading models discussed, the theoretical results in this chapters are automatically verified by these simulations.

4.4 Performance of Coherent MFSK

From (2.110) and (3.32) we may write the expression for average error probability of coherent MFSK in a fading channel as

$$
\begin{aligned}
P_e &= \frac{M-1}{\pi} \int_0^\infty \int_0^{\pi/2} \exp\left(-\frac{\gamma}{2\sin^2\theta}\right) f_\gamma(\gamma) d\theta d\gamma \\
&\quad - \frac{(M-1)(M-2)}{2\pi} \int_0^\infty \int_0^{\pi/4} f_1(\theta,\gamma) f_\gamma(\gamma) d\theta d\gamma \\
&\quad + \frac{(M-1)(M-2)(M-3)}{6\pi^2} \\
&\qquad \times \int_0^\infty \int_0^{\pi/6} f_1(\theta,\gamma) f_2(\theta) f_\gamma(\gamma) d\theta d\gamma \\
&\quad + \frac{(M-1)(M-2)(M-3)}{12\pi^2} \\
&\qquad \times \int_0^\infty \int_0^{\sin^{-1}\left(\frac{1}{\sqrt{3}}\right)} f_1(\theta,\gamma)[\pi - f_2(\theta)] f_\gamma(\gamma) d\theta d\gamma \\
&\quad - \frac{(M-1)(M-2)(M-3)(M-4)}{24\pi^2} \\
&\qquad \times \int_0^\infty \int_0^{\pi/6} f_1(\theta,\gamma) f_2(\theta) f_\gamma(\gamma) d\theta d\gamma
\end{aligned}
\tag{4.63}
$$

Thus, for coherent MFSK our proposed approximation in (4.63) contains a sum of integrals having only exponential functions (with respect to fading SNR γ) as integrands. The double integrals that is to be calculated when the PDF approach is adopted can be reduced to single integrals with the equivalent MGF approach (see (3.34) for details) as

$$
\begin{aligned}
P_e &= \frac{M-1}{\pi} \int_0^{\pi/2} M_\gamma\left(\frac{1}{2\sin^2\theta}\right) d\theta \\
&\quad - \frac{(M-1)(M-2)}{2\pi} \int_0^{\pi/4} \frac{\sin\theta}{\sqrt{1+\sin^2\theta}} M_\gamma\left(\frac{1}{1+\sin^2\theta}\right) d\theta
\end{aligned}
$$

114

$$+ \frac{(M-1)(M-2)(M-3)}{6\pi^2}$$

$$\times \int_0^{\pi/6} \frac{\sin\theta}{\sqrt{1+\sin^2\theta}} M_\gamma\left(\frac{1}{1+\sin^2\theta}\right) f_2(\theta)d\theta$$

$$+ \frac{(M-1)(M-2)(M-3)}{12\pi^2}$$

$$\times \int_0^{\sin^{-1}\left(\frac{1}{\sqrt{3}}\right)} \frac{\sin\theta}{\sqrt{1+\sin^2\theta}} M_\gamma\left(\frac{1}{1+\sin^2\theta}\right) [\pi - f_2(\theta)]d\theta \qquad (4.64)$$

$$- \frac{(M-1)(M-2)(M-3)(M-4)}{24\pi^2}$$

$$\times \int_0^{\pi/6} \frac{\sin\theta}{\sqrt{1+\sin^2\theta}} M_\gamma\left(\frac{1}{1+\sin^2\theta}\right) f_2(\theta)d\theta$$

For deriving (4.64) a change of integral has been performed and the integration over γ has been executed prior to the integration over θ.

For Rayleigh fading the MGF is given by (3.41). Putting this value in (4.64) we can obtain the average error rate as

$$
\begin{aligned}
P_e &= \frac{M-1}{\pi} \int_0^{\pi/2} \frac{2\sin^2\theta}{\bar{\gamma}+2\sin^2\theta} d\theta \\
&\quad - \frac{(M-1)(M-2)}{2\pi} \int_0^{\pi/4} f_3(\theta)d\theta \\
&\quad + \frac{(M-1)(M-2)(M-3)}{6\pi^2} \int_0^{\pi/6} f_3(\theta)f_2(\theta)d\theta \\
&\quad + \frac{(M-1)(M-2)(M-3)}{12\pi^2} \int_0^{\sin^{-1}\left(\frac{1}{\sqrt{3}}\right)} f_3(\theta)[\pi - f_2(\theta)]d\theta \\
&\quad - \frac{(M-1)(M-2)(M-3)(M-4)}{24\pi^2} \int_0^{\pi/6} f_3(\theta)f_2(\theta)d\theta
\end{aligned}
\qquad (4.65)
$$

where

$$f_3(\theta) = \frac{\sin\theta\sqrt{1+\sin^2\theta}}{\bar{\gamma}+1+\sin^2\theta} \qquad (4.66)$$

The same result also can be obtained directly (PDF method) by performing term by term integration of the Rayleigh PDF. The averaging operation involves integration of simple exponential terms in the form of $\int_0^\infty \exp(-ax)dx = 1/a$.

In Fig. 4.5, SEP of coherent MFSK over Rayleigh channel given by (4.65) is depicted for various values of modulation order M (=2, 4, and 10). As our proposed method

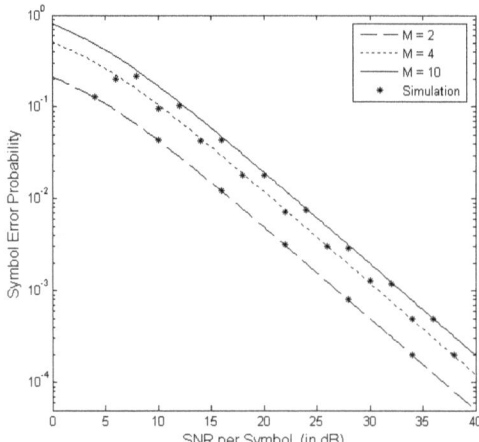

Figure 4.5: SEP of coherent MFSK in Rayleigh fading.

gives exact results up to $M = 5$, the plots for $M = 2$ and $M = 4$ are free from any approximation. The plot for $M = 10$ is supposed to show the inaccuracy due to the approximation process but it is hard to find any such effect, at least with bare eyes. The general characteristics denote that at a particular SNR value, error probability rises with constellation size M. Also all the SEP curves follow almost linear decrease especially at high SNR values.

4.4.1 SEP for MRC

Calculation of the PDF of output SNR for a maximal ratio combiner often becomes cumbersome whereas, evaluation of corresponding MGFs $M_{\gamma_{MRC}}(s) = \int_0^\infty f_{\gamma_{MRC}}(\gamma)\exp(-s\gamma)d\gamma$ are rather easy. The exponential term in the integrand may be decomposed to a product of L different exponentials as $\gamma = \gamma_1 + \gamma_2 + \cdots + \gamma_L$. Further, if the branches are assumed to be independent then the PDF $f_{\gamma_{MRC}}(\gamma)$ also can be expressed as a product of individual PDFs $f_{\gamma_k}(\gamma); k \in \{1, 2, \cdots, L\}$ of L random variables. Thus the single integral becomes an L-fold integral, $M_{\gamma_{MRC}}(s) = \int_0^\infty f_{\gamma_1}(\gamma_1)\exp(-s\gamma_1)d\gamma_1 \cdots \int_0^\infty f_{\gamma_L}(\gamma_L)\exp(-s\gamma_L)d\gamma_L$, where each integral gives the MGF of the corresponding branch, $M_{\gamma_{MRC}}(s) = \prod_{k=1}^{L} M_{\gamma_k}(s)$. Now if we assume that all branches, apart from being independent are identical too (IID assumption), then the MGF with MRC diversity simply becomes the MGF of no-diversity

116

case raised to a power equal to the number of diversity branches (L)

$$M_{\gamma_{MRC}}(s) = \prod_{k=1}^{L} M_{\gamma_k}(s) = [M_\gamma(s)]^L \tag{4.67}$$

where the MGF for Rayleigh fading $M_\gamma(s)$ is defined in (3.41).

Now for SEP evaluation with MRC, we start with (4.67) and by modifying (4.64) we can write the following general equation

$$
\begin{aligned}
P_e &= \frac{M-1}{\pi} \int_0^{\pi/2} \left[M_\gamma \left(\frac{1}{2\sin^2\theta} \right) \right]^L d\theta \\
&\quad - \frac{(M-1)(M-2)}{2\pi} \int_0^{\pi/4} \frac{\sin\theta}{\sqrt{1+\sin^2\theta}} \\
&\qquad\qquad \times \left[M_\gamma \left(\frac{1}{1+\sin^2\theta} \right) \right]^L d\theta \\
&\quad + \frac{(M-1)(M-2)(M-3)}{6\pi^2} \int_0^{\pi/6} \frac{\sin\theta}{\sqrt{1+\sin^2\theta}} \\
&\qquad\qquad \times \left[M_\gamma \left(\frac{1}{1+\sin^2\theta} \right) \right]^L f_2(\theta) d\theta \\
&\quad + \frac{(M-1)(M-2)(M-3)}{12\pi^2} \int_0^{\sin^{-1}\left(\frac{1}{\sqrt{3}}\right)} \frac{\sin\theta}{\sqrt{1+\sin^2\theta}} \\
&\qquad\qquad \times \left[M_\gamma \left(\frac{1}{1+\sin^2\theta} \right) \right]^L [\pi - f_2(\theta)] d\theta \\
&\quad - \frac{(M-1)(M-2)(M-3)(M-4)}{24\pi^2} \int_0^{\pi/6} \frac{\sin\theta}{\sqrt{1+\sin^2\theta}} \\
&\qquad\qquad \times \left[M_\gamma \left(\frac{1}{1+\sin^2\theta} \right) \right]^L f_2(\theta) d\theta
\end{aligned}
\tag{4.68}
$$

to calculate SEP performance over different fading channels for a coherent MFSK receiver employing MRC diversity.

Inserting the MGF from (3.41) in (4.68) we obtain the error performance of a MRC diversity receiver with coherent MFSK in Rayleigh channel as

$$
\begin{aligned}
P_e &= \frac{M-1}{\pi} \int_0^{\pi/2} g_1(\theta) d\theta \\
&\quad - \frac{(M-1)(M-2)}{2\pi} \int_0^{\pi/4} g_2(\theta) d\theta \\
&\quad + \frac{(M-1)(M-2)(M-3)}{6\pi^2} \int_0^{\pi/6} g_2(\theta) f_2(\theta) d\theta \\
&\quad + \frac{(M-1)(M-2)(M-3)}{12\pi^2} \int_0^{\sin^{-1}\left(\frac{1}{\sqrt{3}}\right)} g_2(\theta)[\pi - f_2(\theta)] d\theta \\
&\quad - \frac{(M-1)(M-2)(M-3)(M-4)}{24\pi^2} \int_0^{\pi/6} g_2(\theta) f_2(\theta) d\theta
\end{aligned}
\tag{4.69}
$$

where

$$
g_1(\theta) = \left[\frac{2\sin^2\theta}{\bar{\gamma} + 2\sin^2\theta} \right]^L
\tag{4.70}
$$

and

$$
g_2(\theta) = \frac{\sin\theta}{\sqrt{1+\sin^2\theta}} \left[\frac{1+\sin^2\theta}{\bar{\gamma}+1+\sin^2\theta} \right]^L
\tag{4.71}
$$

As expected, by putting $L = 1$ in (4.69) we get back (4.65). The equations (4.68)-(4.69) demonstrate that MGF approach is quite useful for extending the single-channel SEP expressions to multi-channel reception without much effort.

4.4.2 SEP for SC

It is interesting to note that the method may be advantageous to other diversity combining techniques too. For L branch SC over IID Rayleigh fading channels, the PDF of instantaneous SNR at combiner output can be expressed as

$$
f_{\gamma_{SC}}(\gamma) = \sum_{k=1}^{L} \binom{L}{k} (-1)^{k-1} \frac{k}{\bar{\gamma}} \exp\left(-\frac{k\gamma}{\bar{\gamma}}\right)
\tag{4.72}
$$

which was obtained by differentiating (4.14) and after some rearrangement of terms. The corresponding MGF is

$$
M_{\gamma_{SC}}(s) = \sum_{k=1}^{L} \binom{L}{k} (-1)^{k-1} \frac{k}{k+s\bar{\gamma}}
\tag{4.73}
$$

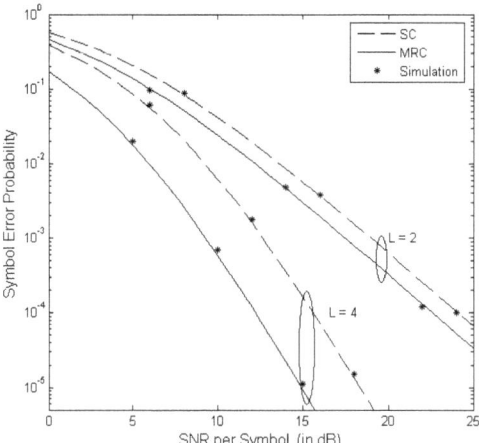

Figure 4.6: SEP of coherent MFSK ($M = 8$) in Rayleigh fading with MRC and SC.

Now from (4.64) and (4.73) it is easy to find the average error rate as

$$
\begin{aligned}
P_e &= \sum_{k=1}^{L} \binom{L}{k}(-1)^{k-1}\left\{ \frac{M-1}{\pi} \int_0^{\pi/2} g_{1,SC}(\theta)d\theta \right. \\
&\quad - \frac{(M-1)(M-2)}{2\pi} \int_0^{\pi/4} g_{2,SC}(\theta)d\theta \\
&\quad + \frac{(M-1)(M-2)(M-3)}{6\pi^2} \int_0^{\pi/6} g_{2,SC}(\theta)f_2(\theta)d\theta \\
&\quad + \frac{(M-1)(M-2)(M-3)}{12\pi^2} \int_0^{\sin^{-1}\left(\frac{1}{\sqrt{3}}\right)} g_{2,SC}(\theta)[\pi - f_2(\theta)]d\theta \\
&\quad \left. - \frac{(M-1)(M-2)(M-3)(M-4)}{24\pi^2} \int_0^{\pi/6} g_{2,SC}(\theta)f_2(\theta)d\theta \right\}
\end{aligned}
\tag{4.74}
$$

where

$$
g_{1,SC}(\theta) = \frac{2k\sin^2\theta}{\bar{\gamma} + 2k\sin^2\theta}
\tag{4.75}
$$

and

$$
g_{2,SC}(\theta) = \frac{k\sin\theta\sqrt{1+\sin^2\theta}}{\bar{\gamma} + k(1+\sin^2\theta)}
\tag{4.76}
$$

The coherent MFSK error rates over Rayleigh fading channel with SC and MRC diversity for various diversity orders ($L = 2$ and 4) is shown in Fig. 4.6, for a fixed constellation size ($M = 8$). The pattern of SEP variation with modulation order M is studied previously (refer to Fig. 4.5) and hence are not repeated here.

4.5 Performance of MQAM

4.5.1 Related Work

Previous related works concerning the error performance of MQAM in conjunction with diversity combining schemes are stated as follows. Lu et al. [64] considered MRC and SC diversity reception and adopted the approach of Chennakeshu [118] to evaluate the SEP of MQAM over Rayleigh fading channel in terms of finite summations of hypergeometric functions. The results are somewhat generalized forms of the solutions presented in an earlier work by Kim et al. [63]. In a separate paper [62], Kim also considered the problem of deriving the BEP of MQAM with L-branch MRC diversity reception in Rayleigh fading using a series expansion of erfc($\cdot$), and the BEP expressions obtained are in the form of an infinite series rather than in closed-form. For EGC, in [65] an approach to evaluate error rates of MQAM operating in Nakagami (Rayleigh is a special case for $m = 1$) fading environment was presented by transforming the average error integral into the frequency domain. Recently, a parallel moments-based approach was presented by Karagiannidis et al. [69], approximating the MGF of the output SNR using Padé's method [119] to analyze the error performance of MQAM for the EGC receiver operating over generalized fading environments.

4.5.2 SEP for SC

To obtain average SEP for multi-channel selection combining of MQAM signals over Rayleigh fading channel, we prefer the CDF method described in Section 4.2.1. For SC the CDF $F_{\gamma_{SC}}(\gamma_b)$ in a favourable form is given by (4.14), whereas $P_e'(\gamma_b)$ can be obtained by differentiating (2.146) using $d/dz[\text{erfc}(z)] = -(2/\sqrt{\pi})\exp(-z^2)$,

$$P_e'(\gamma_b) = -2q\sqrt{\frac{p}{\pi\gamma_b}}\exp(-p\gamma_b) + 2q^2\sqrt{\frac{p}{\pi\gamma_b}}\exp(-p\gamma_b)\text{erfc}(\sqrt{p\gamma_b}) \qquad (4.77)$$

Hence inserting (4.14) and (4.77) in (4.3)

$$P_e = -\int_0^\infty P_e'(\gamma_b)F_{\gamma_{SC}}(\gamma_b)d\gamma_b = \mathcal{I}_1 - \mathcal{I}_2 \qquad (4.78)$$

where

$$\begin{aligned}
\mathcal{I}_1 &= 2q\sqrt{\frac{p}{\pi}}\sum_{k=0}^{L}\binom{L}{k}(-1)^k\int_0^{\infty}\gamma_b^{-1/2}\exp\left[-\gamma_b\left(p+\frac{k}{\bar{\gamma}_b}\right)\right]d\gamma_b \\
&= \sum_{k=0}^{L}\binom{L}{k}(-1)^k 2q\sqrt{\frac{p\bar{\gamma}_b}{k+p\bar{\gamma}_b}}
\end{aligned} \tag{4.79}$$

and

$$\begin{aligned}
\mathcal{I}_2 &= 2q^2\sqrt{\frac{p}{\pi}}\sum_{k=0}^{L}\binom{L}{k}(-1)^k \\
&\quad \times\int_0^{\infty}\gamma_b^{-1/2}\exp\left[-\gamma_b\left(p+\frac{k}{\bar{\gamma}_b}\right)\right]\operatorname{erfc}(\sqrt{p\gamma_b})d\gamma_b \\
&= \frac{4q^2}{\pi}\sum_{k=0}^{L}\binom{L}{k}(-1)^k\sqrt{\frac{p\bar{\gamma}_b}{k+p\bar{\gamma}_b}}\tan^{-1}\left(\sqrt{\frac{k+p\bar{\gamma}_b}{p\bar{\gamma}_b}}\right)
\end{aligned} \tag{4.80}$$

where [8, (6.285.1)] has been used to derive the last integral. Now inserting the expressions of $\mathcal{I}_1$ and $\mathcal{I}_2$, as obtained above in (4.79) and (4.80), in the final error expression (4.78)

$$P_e = \sum_{k=0}^{L}\binom{L}{k}(-1)^k\sqrt{\frac{p\bar{\gamma}_b}{k+p\bar{\gamma}_b}}\left[2q-\frac{4q^2}{\pi}\tan^{-1}\left(\sqrt{\frac{k+p\bar{\gamma}_b}{p\bar{\gamma}_b}}\right)\right] \tag{4.81}$$

For single channel reception, inserting $L=1$ in (4.81) we get

$$P_e = q(2-q) - 2q\sqrt{\frac{p\bar{\gamma}_b}{1+p\bar{\gamma}_b}}\left[1-\frac{2q}{\pi}\tan^{-1}\left(\sqrt{\frac{k+p\bar{\gamma}_b}{p\gamma_b}}\right)\right] \tag{4.82}$$

The expression derived in (4.81) is basically an alternate form of average SEP expression of MQAM with SC over Rayleigh channels as calculated by Kim et al [63],

$$P_e = \frac{\sqrt{M}-1}{M}\left[\left(\sqrt{M}+1\right)-2\mathcal{I}_3-\left(\sqrt{M}-1\right)\mathcal{I}_4\right] \tag{4.83}$$

where

$$\mathcal{I}_3 = L\sum_{k=0}^{L-1}\binom{L-1}{k}(-1)^k\frac{1}{1+k}\sqrt{\frac{p\bar{\gamma}_b}{1+k+p\bar{\gamma}_b}} \tag{4.84}$$

121

and

$$\mathcal{I}_4 = \frac{4L}{\pi} \sum_{k=0}^{L-1} \binom{L-1}{k} (-1)^k \frac{1}{1+k}$$
$$\times \sqrt{\frac{p\bar{\gamma}_b}{1+k+p\bar{\gamma}_b}} \tan^{-1}\left(\sqrt{\frac{p\bar{\gamma}_b}{1+k+p\bar{\gamma}_b}}\right) \quad (4.85)$$

It may be noted that the original expression of $\mathcal{I}_4$ contains some typographical error which is corrected here. Also (4.81) is much more compact than (4.83). The equivalence of the two expressions given by (4.81) and (4.83) can be shown easily by expressing (4.83) in the following form

$$P_e = (2q - q^2) - 2q(1-q)\mathcal{I}_3 - q^2\mathcal{I}_4 \quad (4.86)$$

and expressing $\mathcal{I}_3$ and $\mathcal{I}_4$ as

$$\mathcal{I}_3 = -\sum_{k=1}^{L} \binom{L}{k} (-1)^k \sqrt{\frac{p\bar{\gamma}_b}{k+p\bar{\gamma}_b}} \quad (4.87)$$

$$\mathcal{I}_4 = -2\sum_{k=1}^{L} \binom{L}{k} (-1)^k \sqrt{\frac{p\bar{\gamma}_b}{k+p\bar{\gamma}_b}}$$
$$+ \frac{4}{\pi}\sum_{k=1}^{L} \binom{L}{k} (-1)^k \sqrt{\frac{p\bar{\gamma}_b}{k+p\bar{\gamma}_b}} \tan^{-1}\left(\sqrt{\frac{k+p\bar{\gamma}_b}{p\bar{\gamma}_b}}\right) \quad (4.88)$$

by rearranging some terms and using the relation $\tan^{-1}(x) = (\pi/2) - \tan^{-1}(1/x)$.

4.5.3 SEP for EGC

Ascertaining the lack of a simple and unified framework for MQAM error performance when the receiver employs EGC diversity, this subsection addresses the problem using the classical direct PDF based approach. The performance analysis of EGC is limited to Rayleigh fading and second-order diversity because a closed-form expression for the PDF of a sum of Rayleigh-distributed RVs does not exist for $L > 2$. The newly derived expressions are composed of simple elementary functions.

Utilizing the relation $Q(z) = (1/2)\mathrm{erfc}(z/\sqrt{2}) = (1/2)[1 - \mathrm{erf}(z/\sqrt{2})]$ we first express the CDF of the EGC combiner output SNR given in (3.93) as

$$F_{\gamma_{EGC}}(\gamma) = 1 - \exp\left(-\frac{2\gamma}{\bar{\gamma}}\right) - \sqrt{\frac{\pi\gamma}{\bar{\gamma}}} \exp\left(-\frac{\gamma}{\bar{\gamma}}\right) \mathrm{erf}\left(\sqrt{\frac{\gamma}{\bar{\gamma}}}\right) \quad (4.89)$$

122

Now substituting (4.77) and (4.89) in (4.3) we get

$$P_e = -\int_0^\infty P_e'(\gamma_b) F_{\gamma_{EGC}}(\gamma_b) d\gamma_b = 2q\sqrt{\frac{p}{\pi}} \mathcal{I}_1 - 2q^2\sqrt{\frac{p}{\pi}} \mathcal{I}_2 \qquad (4.90)$$

where

$$\mathcal{I}_1 = \int_0^\infty \frac{1}{\sqrt{\gamma_b}} \exp(-p\gamma_b) F_{\gamma_{EGC}}(\gamma_b) d\gamma_b \qquad (4.91)$$

and

$$\mathcal{I}_2 = \int_0^\infty \frac{1}{\sqrt{\gamma_b}} \exp(-p\gamma_b) \mathrm{erfc}\left(\sqrt{p\gamma_b}\right) F_{\gamma_{EGC}}(\gamma_b) d\gamma_b \qquad (4.92)$$

Inserting the CDF expression given by (4.89) in (4.91), the integral $\mathcal{I}_1$ can be decomposed into three parts

$$\mathcal{I}_1 = \mathcal{I}_{11} - \mathcal{I}_{12} - \mathcal{I}_{13} \qquad (4.93)$$

where

$$\mathcal{I}_{11} = \int_0^\infty \frac{1}{\sqrt{\gamma_b}} \exp(-p\gamma_b) d\gamma_b \qquad (4.94)$$

$$\mathcal{I}_{12} = \int_0^\infty \frac{1}{\sqrt{\gamma_b}} \exp\left[-\left(p + \frac{2}{\bar{\gamma}_b}\right)\gamma_b\right] d\gamma_b \qquad (4.95)$$

$$\mathcal{I}_{13} = \sqrt{\frac{\pi}{\bar{\gamma}_b}} \int_0^\infty \exp\left[-\left(p + \frac{1}{\bar{\gamma}_b}\right)\gamma_b\right] \mathrm{erf}\left(\sqrt{\frac{\gamma_b}{\bar{\gamma}_b}}\right) d\gamma_b \qquad (4.96)$$

Using $\int_0^\infty z^{n-1} \exp(-\beta z) dz = \Gamma(n)/\beta^n$ and noting that $\Gamma(1/2) = \sqrt{\pi}$, the first two components $\mathcal{I}_{11}$ and $\mathcal{I}_{12}$ can be evaluated as

$$\mathcal{I}_{11} = \sqrt{\frac{\pi}{p}} \qquad (4.97)$$

$$\mathcal{I}_{12} = \sqrt{\frac{\pi\bar{\gamma}_b}{2 + p\bar{\gamma}_b}} \qquad (4.98)$$

and for the third component $\mathcal{I}_{13}$, the integral is solved using [9, (7.4.19)]

$$\mathcal{I}_{13} = \frac{1}{1 + p\bar{\gamma}_b}\sqrt{\frac{\pi\bar{\gamma}_b}{2 + p\bar{\gamma}_b}} \qquad (4.99)$$

Inserting expressions for $\mathcal{I}_{11}$, $\mathcal{I}_{12}$, and $\mathcal{I}_{13}$ in (4.93), $\mathcal{I}_1$ can be finally expressed as

$$
\begin{aligned}
\mathcal{I}_1 &= \sqrt{\frac{\pi}{p}} - \frac{\sqrt{\pi\bar{\gamma}_b(2 + p\bar{\gamma}_b)}}{1 + p\bar{\gamma}_b} \\
&= \sqrt{\frac{\pi}{p}}\left[1 - \sqrt{1 - \frac{1}{(1 + p\bar{\gamma}_b)^2}}\right]
\end{aligned}
\tag{4.100}
$$

The presence of $\mathrm{erfc}^2(\cdot)$ term in the SEP expression (2.146) makes the integral I_2 difficult to evaluate and it is a common practice to drop the second term for simplification. However when SNR $\bar{\gamma}_b$ is low, the term has noticeable effect on average SEP [65]. In addition, the approximation becomes quite crude when the constellation size (M) becomes larger. To evaluate (4.92) we again express I_2 in terms of three components

$$
I_2 = \mathcal{I}_{21} - \mathcal{I}_{22} - \mathcal{I}_{23}
\tag{4.101}
$$

where

$$
\mathcal{I}_{21} = \int_0^\infty \frac{1}{\sqrt{\gamma_b}} \exp(-p\gamma_b)\mathrm{erfc}\left(\sqrt{p\gamma_b}\right) d\gamma_b
\tag{4.102}
$$

$$
\mathcal{I}_{22} = \int_0^\infty \frac{1}{\sqrt{\gamma_b}} \exp\left[-\left(p + \frac{2}{\bar{\gamma}_b}\right)\gamma_b\right] \mathrm{erfc}\left(\sqrt{p\gamma_b}\right) d\gamma_b
\tag{4.103}
$$

$$
\mathcal{I}_{23} = \sqrt{\frac{\pi}{\bar{\gamma}_b}} \int_0^\infty \exp\left[-\left(p + \frac{1}{\bar{\gamma}_b}\right)\gamma_b\right] \mathrm{erfc}\left(\sqrt{p\gamma_b}\right) \mathrm{erf}\left(\sqrt{\frac{\gamma_b}{\bar{\gamma}_b}}\right) d\gamma_b
\tag{4.104}
$$

which results when (4.89) is substituted in (4.92). Now, using [8, (6.285.1)] $\mathcal{I}_{21}$ and $\mathcal{I}_{22}$ can be readily evaluated to

$$
\mathcal{I}_{21} = \frac{1}{2}\sqrt{\frac{\pi}{p}}
\tag{4.105}
$$

$$
\mathcal{I}_{22} = \frac{2}{\sqrt{\pi}}\sqrt{\frac{\bar{\gamma}_b}{2 + p\bar{\gamma}_b}} \tan^{-1}\left(\sqrt{\frac{2 + p\bar{\gamma}_b}{p\bar{\gamma}_b}}\right)
\tag{4.106}
$$

whereas using the relation $\mathrm{erfc}(z) = 1 - \mathrm{erf}(z)$ we can express $\mathcal{I}_{23}$ as

$$
\mathcal{I}_{23} = \sqrt{\frac{\pi}{\bar{\gamma}_b}} [\mathcal{I}_{231} - \mathcal{I}_{232}]
\tag{4.107}
$$

where

$$
\mathcal{I}_{231} = \int_0^\infty \exp\left[-\left(p + \frac{1}{\bar{\gamma}_b}\right)\gamma_b\right] \mathrm{erf}\left(\sqrt{\frac{\gamma_b}{\bar{\gamma}_b}}\right) d\gamma_b
\tag{4.108}
$$

and

$$\mathcal{I}_{232} = \int_0^\infty \exp\left[-\left(p + \frac{1}{\bar{\gamma}_b}\right)\gamma_b\right] \operatorname{erf}\left(\sqrt{p\gamma_b}\right) \operatorname{erf}\left(\sqrt{\frac{\gamma_b}{\bar{\gamma}_b}}\right) d\gamma_b \tag{4.109}$$

The first term can be expressed in closed-form using [9, (7.4.19)]

$$\mathcal{I}_{231} = \frac{\bar{\gamma}_b}{1 + p\bar{\gamma}_b}\sqrt{\frac{1}{2 + p\bar{\gamma}_b}} \tag{4.110}$$

while the integral in second term can be evaluated using [76, (App.A.10)]

$$\mathcal{I}_{232} = \frac{2\bar{\gamma}_b}{\pi(1 + p\bar{\gamma}_b)}\left[\sqrt{\frac{p\bar{\gamma}_b}{1 + 2p\bar{\gamma}_b}}\tan^{-1}\left(\sqrt{\frac{1}{1 + 2p\bar{\gamma}_b}}\right)\right.$$
$$\left. + \sqrt{\frac{1}{2 + p\bar{\gamma}_b}}\tan^{-1}\left(\sqrt{\frac{p\bar{\gamma}_b}{2 + p\bar{\gamma}_b}}\right)\right] \tag{4.111}$$

Inserting the expressions for $\mathcal{I}_{231}$ and $\mathcal{I}_{232}$ from (4.110) and (4.111) respectively in (4.107) we have

$$\mathcal{I}_{23} = \frac{1}{1 + p\bar{\gamma}_b}\sqrt{\frac{\pi\bar{\gamma}_b}{2 + p\bar{\gamma}_b}} - \frac{2}{1 + p\bar{\gamma}_b}\sqrt{\frac{\bar{\gamma}_b}{\pi}}$$
$$\times \left[\sqrt{\frac{p\bar{\gamma}_b}{1 + 2p\bar{\gamma}_b}}\tan^{-1}\left(\sqrt{\frac{1}{1 + 2p\bar{\gamma}_b}}\right)\right.$$
$$\left. + \sqrt{\frac{1}{2 + p\bar{\gamma}_b}}\tan^{-1}\left(\sqrt{\frac{p\bar{\gamma}_b}{2 + p\bar{\gamma}_b}}\right)\right] \tag{4.112}$$

and from (4.105), (4.106), and (4.112)

$$\mathcal{I}_2 = \frac{1}{2}\sqrt{\frac{\pi}{p}} - \frac{\sqrt{\pi\bar{\gamma}_b(2 + p\bar{\gamma}_b)}}{1 + p\bar{\gamma}_b}$$
$$+ \frac{2}{1 + p\bar{\gamma}_b}\sqrt{\frac{\bar{\gamma}_b(2 + p\bar{\gamma}_b)}{\pi}}\tan^{-1}\left(\sqrt{\frac{p\bar{\gamma}_b}{2 + p\bar{\gamma}_b}}\right)$$
$$+ \frac{2\bar{\gamma}_b}{1 + p\bar{\gamma}_b}\sqrt{\frac{p}{\pi(1 + 2p\bar{\gamma}_b)}}\tan^{-1}\left(\sqrt{\frac{1}{1 + 2p\bar{\gamma}_b}}\right) \tag{4.113}$$

Substituting (4.100) and (4.113) in (4.90), closed form expression for average SEP of square MQAM with two-branch EGC over Rayleigh fading channel becomes

$$
\begin{aligned}
P_e = & \left(2q - q^2\right) + \frac{2q(q-1)}{1 + p\bar{\gamma}_b}\sqrt{p\bar{\gamma}_b(2 + p\bar{\gamma}_b)} \\
& - \frac{4q^2\sqrt{p\bar{\gamma}_b}}{\pi(1 + p\bar{\gamma}_b)}\left[\sqrt{2 + p\bar{\gamma}_b}\tan^{-1}\left(\sqrt{\frac{p\bar{\gamma}_b}{2 + p\bar{\gamma}_b}}\right)\right. \\
& \left. + \sqrt{\frac{p\bar{\gamma}_b}{1 + 2p\bar{\gamma}_b}}\tan^{-1}\left(\sqrt{\frac{1}{1 + 2p\bar{\gamma}_b}}\right)\right]
\end{aligned}
\tag{4.114}
$$

As the in-phase and the quadrature components of square MQAM may be demodulated and detected independently, the conditional SEP expression over AWGN channel given in (2.146) may be written in the following way [68]

$$
P_e(\gamma_b) = 1 - [1 - P_I(\gamma_b)][1 - P_Q(\gamma_b)]
\tag{4.115}
$$

where, $P_I(\gamma_b)$ and $P_Q(\gamma_b)$ are the conditional SEP of the in-phase and the quadrature components. Since $\log_2(M)$ is even, $P_I(\gamma_b)$ and $P_Q(\gamma_b)$ are equal and they are equivalent to the conditional SEP of MPAM with $\sqrt{M}$ signals

$$
P_I(\gamma_b) = P_Q(\gamma_b) = P_{\sqrt{M}-\text{PAM}}(\gamma_b) = q\,\text{erfc}(\sqrt{p\gamma_b})
\tag{4.116}
$$

Thus

$$
\begin{aligned}
P_e(\gamma_b) &= 1 - \left[1 - P_{\sqrt{M}-\text{PAM}}(\gamma_b)\right]^2 \\
&= 2P_{\sqrt{M}-\text{PAM}}(\gamma_b) - P_{\sqrt{M}-\text{PAM}}^2(\gamma_b)
\end{aligned}
\tag{4.117}
$$

Similarly, in case of Rayleigh fading channel, the average SEP of MQAM can be obtained by calculating average SEP for individual MPAM components as

$$
P_e \approx 2P_{\sqrt{M}-\text{PAM}} - P_{\sqrt{M}-\text{PAM}}^2
\tag{4.118}
$$

where $P_{\sqrt{M}-\text{PAM}}$ is given by

$$
P_{\sqrt{M}-\text{PAM}} = q\int_0^\infty \text{erfc}(\sqrt{p\gamma_b})f_{\gamma_{EGC}}(\gamma_b)d\gamma_b
\tag{4.119}
$$

which may be evaluated using the CDF method as

$$
P_{\sqrt{M}-\text{PAM}} = q\sqrt{\frac{p}{\pi}}\int_0^\infty \frac{1}{\sqrt{\gamma_b}}\exp(-p\gamma_b)F_{\gamma_{EGC}}(\gamma_b)d\gamma_b
\tag{4.120}
$$

126

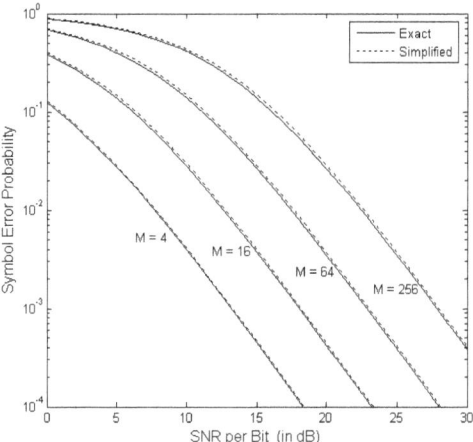

Figure 4.7: SEP of square MQAM with EGC in Rayleigh fading.

Noting the similarity between (4.91) and (4.120) we may write from (4.100)

$$P_{\sqrt{M}-\text{PAM}} = q \left[1 - \sqrt{1 - \frac{1}{(1 + p\bar{\gamma}_b)^2}} \right] \tag{4.121}$$

and thus the simplified SEP expression in (4.118) becomes

$$P_e \approx 2q \left[1 - \sqrt{1 - \frac{1}{(1 + p\bar{\gamma}_b)^2}} \right] - q^2 \left[1 - \sqrt{1 - \frac{1}{(1 + p\bar{\gamma}_b)^2}} \right]^2 \tag{4.122}$$

Average SEP performance of second order EGC diversity employing square MQAM with both the exact and simplified models, as available from (4.114) and (4.122), are depicted in Fig. 4.7 for various constellation size M (= 4, 16, 64, and 256). From the plots one can identify that as the modulation order M increases, for a given average SNR per bit, the error probability also increases demonstrating the power penalty for the sake of bandwidth efficiency.

To study the difference of error probabilities calculated via exact and simplified model, we define the percentage error of simplified SEP as

$$\text{Percentage} \quad \text{Error} = \frac{P_{e,\text{Exact}} - P_{e,\text{Simplified}}}{P_{e,\text{Exact}}} \times 100\% \tag{4.123}$$

where $P_{e,\text{Exact}}$ and $P_{e,\text{Simplified}}$ denotes SEP of square MQAM with dual diversity EGC

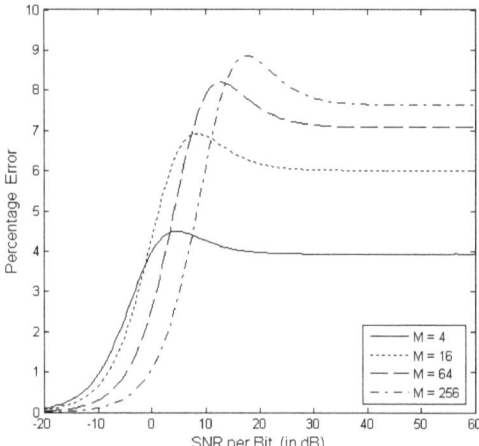

Figure 4.8: Percentage error of the simplified SEP with respect to the exact SEP value for square MQAM with dual diversity EGC in Rayleigh fading.

operating over Rayleigh channel computed through exact method and using simplified method respectively. The derived error is plotted against SNR per bit per channel in Fig. 4.8, which reveals that the discrepancy between the two models is within 4% to 10%. With increasing SNR the error first increases and then saturates for high SNR values. Again, the error for a given SNR is more for higher values of M. It is worthwhile to note from Fig. 4.8 that the average SEP of the simplified model is always higher than or equal to the SEP of the common model, and thereby giving an upper bound.

4.6 Performance of Binary Modulations in Double Rayleigh Fading with Switched Combining

Recent studies reveal that the classical double Rayleigh model has been already extended to models containing the product of two or more Rician [103, 120, 121], Nakagami [122, 123], and Weibull [124] RVs. The references cited here, though by no means comprehensive, highlight the growing research interest about keyhole M2M communication. In spite of this, we find that literature dealing with error performance of different modulation schemes in keyhole channels are still scarce. Exceptions include an approximate solution of probability of error by Salo et al. [125], where the authors addressed coherent modulation schemes (having Gaussian Q function in the error probability expression) and evaluated the error rates in a cascaded Rayleigh channel. In a separate work, Uysal

[126] calculated MPSK error performance in cascaded Rayleigh channels and presented the results in the form of single finite range integral. Some more studies [127, 128, 129] on double Rayleigh fading channel have also been done recently with space-time codes. However, to the best of the authors' knowledge, error performance of different diversity systems in cascaded fading channel has never been investigated.

For BER calculation over a cascaded Rayleigh fading channel we start our derivation with a formal definition of the channel SNR statistics. Let us consider two independent Rayleigh distributed signal envelopes α_1 and α_1 with mean power $E\{\alpha_1^2\} = \Omega_1$ and $E\{\alpha_2^2\} = \Omega_2$ respectively. Following the arguments in Kovacs' thesis [130, (A.2.10)], we can express the PDF of the cascaded signal envelope $\alpha = \alpha_1\alpha_2$, i.e. $f_\alpha(\alpha)$ as

$$f_\alpha(\alpha) = \frac{4\alpha}{\Omega} K_0 \left(\frac{2\alpha}{\sqrt{\Omega}} \right) \qquad ; \alpha \geq 0 \tag{4.124}$$

where $E\{\alpha^2\} = \Omega_1\Omega_2 = \Omega$ and $K_0(\cdot)$ denotes zero order modified Bessel function of second kind. Next, with a simple transformation of variable $f_\gamma(\gamma) = f_\alpha(\sqrt{\gamma\Omega/\bar{\gamma}})/(2\sqrt{\gamma\bar{\gamma}/\Omega})$ (Chapter 2.4, (3.29)), the PDF of instantaneous SNR γ may be obtained as

$$f_\gamma(\gamma) = \frac{2}{\bar{\gamma}} K_0 \left(2\sqrt{\frac{\gamma}{\bar{\gamma}}} \right) \qquad ; \gamma \geq 0 \tag{4.125}$$

which was first mentioned by Chizhik et al. [131, (5)]. Further, integrating (4.125) using Luke's indefinite integration formula [132, (5.2.31)]

$$\int_0^z x^\nu K_{\nu-1}(x)dx = 2^{\nu-1}\Gamma(\nu) - z^\nu K_\nu(z) \qquad ; \Re\{\nu\} > 0 \tag{4.126}$$

one arrives at the corresponding CDF of γ

$$F_\gamma(\gamma) = 1 - 2\sqrt{\frac{\gamma}{\bar{\gamma}}} K_1 \left(2\sqrt{\frac{\gamma}{\bar{\gamma}}} \right) \qquad ; \gamma \geq 0 \tag{4.127}$$

4.6.1 BEP for SEC

It is quite evident from Section 4.2.6 that the performance of SSC diversity receivers can be derived as a special case ($L = 2$) of diversity systems employing SEC. Thus for cascaded Rayleigh fading channels, we deduce the unified BEP for SEC systems first, and then specialize the end expression to obtain BEP for SSC systems in the next subsection.

From (3.108), for arbitrary fading channels, the PDF for L branch SEC would be

$$f_{\gamma_{SEC}}(\gamma_b) = \begin{cases} f_\gamma(\gamma_b)[F_\gamma(\gamma_T)]^{L-1} & ; \gamma_b < \gamma_T \\ f_\gamma(\gamma_b) \sum_{k=0}^{L-1} [F_\gamma(\gamma_T)]^k & ; \gamma_b \geq \gamma_T \end{cases} \tag{4.128}$$

where $f_\gamma(\gamma_b)$, $F_\gamma(\gamma_b)$ are the PDF and CDF of the corresponding fading channel SNR. For cascaded Rayleigh fading channel the PDF and CDF are given by (4.125) and (4.127) respectively.

Averaging the conditional BEP $P_{e,b}(\gamma_b)$ for binary modulations given in (2.42) over (4.128), and after some rearrangement of terms, the average BEP with SEC thus can be found as

$$
\begin{aligned}
P_{e,b} &= \sum_{k=0}^{L-1} [F_\gamma(\gamma_T)]^k \int_0^\infty P_{e,b}(\gamma_b) f_\gamma(\gamma_b) d\gamma_b \\
&\quad - \sum_{k=0}^{L-2} [F_\gamma(\gamma_T)]^k \int_0^{\gamma_T} P_{e,b}(\gamma_b) f_\gamma(\gamma_b) d\gamma_b \\
&= \mathcal{I}_1 - \mathcal{I}_2
\end{aligned}
\tag{4.129}
$$

Utilizing the relation $\partial \Gamma(a, z)/\partial z = -z^{a-1} \exp(-z)$ [9, (6.5.25)] and (B.15), the first integral is evaluated easily through integration by parts

$$
\begin{aligned}
\mathcal{I}_1 &= \sum_{k=0}^{L-1} [F_\gamma(\gamma_T)]^k \\
&\quad \times \frac{1}{2} \left[1 - \frac{\Gamma(\beta+1)}{\alpha \bar{\gamma}_b} U\left(\beta+1; 2; \frac{1}{\alpha \bar{\gamma}_b}\right) \right]
\end{aligned}
\tag{4.130}
$$

where $U(\cdot; \cdot; \cdot)$ stands for Tricomi's confluent hypergeometric function [9, (13.1.3)]. It may be noted that second part of (4.130) basically gives the BEP of different binary modulations in cascaded Rayleigh fading channel (no diversity, $L = 1$) in closed-form, i.e.

$$
\int_0^\infty P_{e,b}(\gamma_b) f_\gamma(\gamma_b) d\gamma_b = \frac{1}{2} \left[1 - \frac{\Gamma(\beta+1)}{\alpha \bar{\gamma}_b} U\left(\beta+1; 2; \frac{1}{\alpha \bar{\gamma}_b}\right) \right]
\tag{4.131}
$$

and thus avoids computation of any infinite series, common in previous literature [125].

Unfortunately a similar closed-form expression can not be found for the second integral which has a finite integration range. That's why for the second integral, first, we represent the conditional error probability $P_{e,b}(\gamma_b)$ in series form [8, (8.354.2)]

$$
P_{e,b}(\gamma_b) = \frac{1}{2} - \frac{\alpha^\beta}{2\Gamma(\beta)} \sum_{j=0}^\infty \frac{(-1)^j \alpha^j}{j!(\beta+j)} \gamma_b^{\beta+j}
\tag{4.132}
$$

and then restate the second integral in (4.129) as

$$
\begin{aligned}
\mathcal{I}_2 &= \frac{1}{2} \sum_{k=0}^{L-1} [F_\gamma(\gamma_T)]^k - \frac{\alpha^\beta}{2\Gamma(\beta)} \sum_{k=0}^{L-2} [F_\gamma(\gamma_T)]^k \sum_{j=0}^\infty \frac{(-1)^j \alpha^j}{j!(\beta+j)} \\
&\quad \times \int_0^{\gamma_T} \gamma_b^{\beta+j} f_\gamma(\gamma_b) d\gamma_b
\end{aligned}
\tag{4.133}
$$

Lastly, using [133, (1.12.1.6)], the integral in (4.133) may be evaluated as follows

$$
\int_0^{\gamma_T} \gamma_b^{\beta+j} f_\gamma(\gamma_b) d\gamma_b = \frac{2\hat{\gamma}_b \gamma_T^{\beta+j}}{\beta+j+1} K_0 \left(2\sqrt{\hat{\gamma}_b}\right)
$$
$$
\times\, _1F_2\left(1; \beta+j+1, \beta+j+2; \hat{\gamma}_b\right)
$$
$$
+\frac{2\hat{\gamma}_b^{3/2} \gamma_T^{\beta+j}}{(\beta+j+1)^2} K_1\left(2\sqrt{\hat{\gamma}_b}\right)
$$
$$
\times\, _1F_2\left(1; \beta+j+2, \beta+j+2; \hat{\gamma}_b\right)
$$

(4.134)

where $_pF_q(\{a_1, \cdots, a_p\}; \{b_1, \cdots, b_q\}; z)$ refers to the generalized hypergeometric function [8, (9.14.1)] and $\hat{\gamma}_b = \gamma_T/\bar{\gamma}_b$ denotes normalized threshold as usual.

Inserting (4.134) and (4.130) in (4.129), and putting everything together, unified BEP expression for binary modulations with SEC in a cascaded Rayleigh fading channel is obtained as

$$
P_{e,b} = \frac{1}{2}\left[1 - \sum_{k=0}^{L-1} [F_\gamma(\gamma_T)]^k \frac{\Gamma(\beta+1)}{\alpha\bar{\gamma}_b} U\left(\beta+1; 2; \frac{1}{\alpha\bar{\gamma}_b}\right)\right]
$$
$$
+\frac{\hat{\gamma}_b}{\Gamma(\beta)} \sum_{k=0}^{L-2} [F_\gamma(\gamma_T)]^k \sum_{j=0}^{\infty} \frac{(-1)^j (\alpha\gamma_T)^{\beta+j}}{j!(\beta+j)(\beta+j+1)}
$$
$$
\times \left[K_0\left(2\sqrt{\hat{\gamma}_b}\right) \, _1F_2\left(1; \beta+j+1, \beta+j+2; \hat{\gamma}_b\right)\right.
$$
$$
+K_1\left(2\sqrt{\hat{\gamma}_b}\right)
$$
$$
\left. \times \frac{\sqrt{\hat{\gamma}_b}}{\beta+j+1} \, _1F_2\left(1; \beta+j+2, \beta+j+2; \hat{\gamma}_b\right)\right]
$$

(4.135)

The final expression involves simple hypergeometric functions which can be evaluated easily and efficiently. Also the series term converges very fast, an accuracy of 10^{-14} can be achieved within first 50 terms.

4.6.2 BEP for SSC

The BEP for SSC case may be obtained by putting $L = 2$ in (4.135) as

$$
\begin{aligned}
P_{e,b} &= \frac{1}{2}\left[1 - \{1 + F_{\gamma}(\gamma_T)\}\frac{\Gamma(\beta+1)}{\alpha\bar{\gamma}_b}U\left(\beta+1;2;\frac{1}{\alpha\bar{\gamma}_b}\right)\right] \\
&+ \frac{\hat{\gamma}_b}{\Gamma(\beta)}\sum_{j=0}^{\infty}\frac{(-1)^j(\alpha\gamma_T)^{\beta+j}}{j!(\beta+j)(\beta+j+1)} \\
&\times \left[K_0\left(2\sqrt{\hat{\gamma}_b}\right) {}_1F_2\left(1;\beta+j+1,\beta+j+2;\hat{\gamma}_b\right)\right. \\
&+ K_1\left(2\sqrt{\hat{\gamma}_b}\right) \\
&\times \left. \frac{\sqrt{\hat{\gamma}_b}}{\beta+j+1} {}_1F_2\left(1;\beta+j+2,\beta+j+2;\hat{\gamma}_b\right)\right]
\end{aligned}
\tag{4.136}
$$

Further, by inserting values of the parameter set $\{\alpha, \beta\}$ in (4.136) one may obtain error rates for different modulation methods. For example, in case of BPSK ($\alpha = 1, \beta = 1/2$) the BEP would be

$$
\begin{aligned}
P_{e,b} &= \frac{1}{2}\left[1 - \{1 + F_{\gamma}(\gamma_T)\}\frac{\exp\left(\frac{1}{2\bar{\gamma}_b}\right)}{2\bar{\gamma}_b}\right. \\
&\times \left.\left\{K_1\left(\frac{1}{2\bar{\gamma}_b}\right) - K_0\left(\frac{1}{2\bar{\gamma}_b}\right)\right\}\right] \\
&+ \frac{\hat{\gamma}_b}{\sqrt{\pi}}\sum_{j=0}^{\infty}\frac{(-1)^j\gamma_T^{\beta+j}}{j!(j+1/2)(j+3/2)} \\
&\times \left[K_0\left(2\sqrt{\hat{\gamma}_b}\right) {}_1F_2\left(1;j+\frac{3}{2},j+\frac{5}{2};\hat{\gamma}_b\right)\right. \\
&+ K_1\left(2\sqrt{\hat{\gamma}_b}\right) \\
&\times \left. \frac{\sqrt{\hat{\gamma}_b}}{j+3/2} {}_1F_2\left(1;j+\frac{5}{2},j+\frac{5}{2};\hat{\gamma}_b\right)\right]
\end{aligned}
\tag{4.137}
$$

where $\Gamma(1/2) = \sqrt{\pi}$ [9, (6.1.8)], $\Gamma(3/2) = \sqrt{\pi}/2$ [9, (6.1.9)], and $U(3/2; 2; x) = (1/\sqrt{\pi})\exp(x/2)[K_1(x/2) K_0(x/2)]$ [134, (07.33.03.0446.01)] have been used to derive (4.137).

For finding the optimal threshold under cascaded fading, we opted for an indirect numerical method as differentiation of (4.135) or (4.136) would generate infinite number of terms. The method finds minimum error probability that ensures optimality of the derived threshold value. The threshold values are made accurate to two decimal places. Detailed results for BPSK and DPSK modulation that were found through this procedure are depicted in Table 4.2. From Table 4.2 one finds that the optimal values follow a trend similar to single Rayleigh fading, i.e. the values are increasing with both $\bar{\gamma}_b$ and L.

$\bar{\gamma}_b$ (dB)	Optimum common switching threshold γ_T^* (dB)					
	BPSK ($\alpha = 1$, $\beta = \frac{1}{2}$)			DPSK($\alpha = 1$, $\beta = 1$)		
	L=8	L=4	SSC	L=8	L=4	SSC
0	-0.04	-2.09	-4.44	0.80	-0.95	-2.87
3	1.93	-0.16	-2.56	2.67	0.86	-1.12
6	3.68	1.53	-0.93	4.33	2.44	0.38
9	5.25	3.01	0.47	5.82	3.83	1.66
12	6.65	4.32	1.67	7.14	5.05	2.76
15	7.90	5.47	2.72	8.34	6.13	3.72
18	9.03	6.50	3.63	9.42	7.10	4.55
21	10.03	7.42	4.43	10.38	7.96	5.29
24	10.94	8.24	5.13	11.26	8.74	5.94
27	11.71	8.99	5.77	11.79	9.44	6.52

Table 4.2: Optimum switching threshold γ_T^* for BPSK and DPSK with SEC in double Rayleigh fading.

Nevertheless, for a given average SNR and diversity order, γ_T^* values are lesser for the cascaded channel no matter whatever modulation scheme is in use.

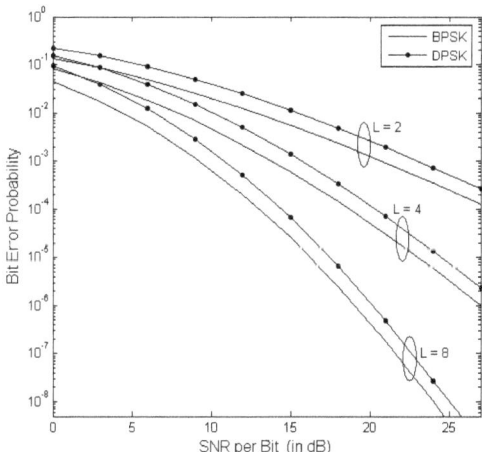

Figure 4.9: BEP of BPSK and DPSK with SSC/ SEC in double Rayleigh fading.

In Fig. 4.9, we present the average BEP of SSC and SEC diversity receivers operating over a cascaded Rayleigh fading channel. Error rate curves for different diversity orders (L = 2, 4, and 8) are shown. For calculating BEP at each SNR point ($\bar{\gamma}_b$), optimal threshold values given in Table 4.2 are used.

133

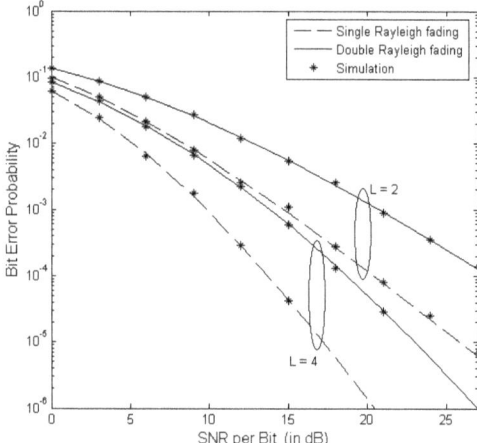

Figure 4.10: Analytical and simulated BEP of BPSK with SSC/SEC ($L = 2$ and 4) in single and double Rayleigh fading.

We compare the performance of switched combining in cascaded Rayleigh channel with respect to the simple Rayleigh channel in Fig. 4.10. The negative effect of keyhole channels is clear. Keeping the available SNR at receiver constant, for any particular modulation, if the channel characteristics change from simple Rayleigh to cascaded Rayleigh the error rates may go up at least an order of magnitude. This may result in irreducible error floor in some QoS limited systems, and calls for additional fading mitigation endeavours (increasing diversity branches, channel coding etc.).

Chapter Summary

In the present chapter, a unified approach is presented for calculation of error probability of binary modulations over both single and double Rayleigh fading channels with different diversity combining techniques. For single Rayleigh fading channel, unified BEP expressions are provided for SC, EGC, and MRC in terms of the modulation parameters α and β. In case of switched diversity, the end expressions also contain the switching threshold SNR, γ_T. Using numerical techniques, optimum switching threshold (γ_T^*) values have been computed for different channel SNRs ($\bar{\gamma}_b$), and are listed in a tabular form. A similar exercise is carried out for the double Rayleigh fading channel to find the corresponding BEP improvement, when switched diversity is applied. It is found that the degree of improvement was more for double Rayleigh fading channel.

In addition, we have also studied the error performance for other M-ary modulations namely, $\pi/4$-DQPSK, coherent MFSK, and MQAM over single Rayleigh fading channel, in presence of diversity. Simple closed-form expressions for BEP of $\pi/4$-DQPSK with SC, EGC, MRC, and SSC are found, and an improved SEP approximation for the coherent MFSK with MRC is presented. For MQAM, we computed the SEP with SC and EGC following CDF approach. The CDF method is a powerful alternative to the popular PDF approach, and often results in simpler forms. A simple approximation for the SEP with EGC case is also provided and the approximated values are compared graphically with the exact SEP values. The percentage of error is negligible and becomes constant at higher SNR ranges.

Chapter 5

Diversity Combining in Rician Fading

Cellular architecture, which was introduced almost 30 years back, is being scaled down every other day to keep pace with the increasing demand of capacity and reliability. The geographical area covered by a cell is diminishing from tens of kilometers to a few meters as we move from macrocell to microcell to picocell and further to femtocellular domain [135]. In contrast to conventional cells, these pico or femto cells operate with low-power, low-height antennas, which provide LOS path between BS/ AP and mobile units. The receiver is generally surrounded by many obstacles, especially in the urban area. In such a multipath environment, the fading amplitude of the received signal follows a Rician PDF rather than Rayleigh fading, as Rayleigh fading is often a feature of large cells with no direct LOS path [21, 136]. In addition, a parallel development by WiMAX forum (IEEE 802.16j task group) to introduce mobile multihop relaying [137], opens up new possibilities of LOS communication among mobile terminals and nearest relay nodes. Rician fading may also be used to model long-range radio channels like suburban land mobile system [138], maritime satellite links [139] and satellite mobile channels [140] as well as fixed indoor wireless solutions like wireless local area network (WLAN) [141], cordless telecommunication (CT) [21] and factory aisles [142]. Experimental measurements performed at different frequency bands, e.g. for terrestrial mobile systems at 900 MHz [143] and for indoor wireless channels at 1.9 GHz [144], 2.4 GHz [145] and 5 GHz [146] also confirm that the temporal envelope fading is best fit by Rician distribution.

In this chapter, we begin with a brief review of BEP calculation for binary modulations in Section 5.1 and then discuss the performance of $\pi/4$-DQPSK with MRC in Section 5.2. This is followed by SEP analysis of coherent and non-coherent MFSK, in Section 5.3 and Section 5.4 respectively. Coherent optimal combining scheme, i.e. MRC is chosen for the coherent MFSK case, while the SEP for ncMFSK is evaluated with SC as it does not require a phase coherent receiver. Next, in Section 5.5, we present a unified approach for SEP calculation for coherent modulation schemes in Rician fading channel. SEP for several binary, quaternary, and M-ary modulations like, square MQAM, BPSK, BFSK, DBPSK, QPSK, OQPSK, and MSK may be found easily with the generalized approach.

5.1 Performance of Binary Modulations

Among the previous related work on Rician channel, Lindsey [147] demonstrated the performance of binary coherent multireceiver and the non-coherent multireceiver, while Roberts and Bargallo [148] derived an expression for the BEP of DPSK and ncBFSK. The results had been extended to the M-ary case without diversity [21] and with MRC diversity [149] by Sun et al. The probability of error for coherent detection over Rician channel in closed-form is given hereunder [147]

$$P_{e,b} = Q_1(u,v) - \frac{1}{2}\left[1 + \sqrt{\frac{\alpha\bar{\gamma}_b}{1 + K + \alpha\bar{\gamma}_b}}\right] \exp\left(-\frac{u^2 + v^2}{2}\right) I_0(uv) \tag{5.1}$$

where

$$u = \sqrt{y\left(\frac{\sqrt{1 - B^2} - 1}{B}\right)}, \quad v = \sqrt{y\left(\frac{B}{\sqrt{1 - B^2} - 1}\right)} \tag{5.2}$$

are defined in terms of y and B

$$y = \frac{K(1 + K)}{2(1 + K + \alpha\bar{\gamma}_b)}, \quad B = -\frac{1 + K}{1 + K + 2\alpha\bar{\gamma}_b} \tag{5.3}$$

while $\alpha = 1$ for BPSK and $\alpha = 1/2$ for BFSK. For the non coherent modulations the corresponding forms are [148]

$$P_{e,b} = \frac{1 + K}{2(1 + K + \alpha\bar{\gamma}_b)} \exp\left(-\frac{\alpha K \bar{\gamma}_b}{1 + K + \alpha\bar{\gamma}_b}\right) \tag{5.4}$$

where $\alpha = 1$ for DPSK and $\alpha = 1/2$ for ncBFSK. The expression in (5.4) is obtained (with η_b replaced by γ_b) through averaging (2.47) and (2.48) over the Rician PDF given in (3.49) and using (B.12).

A unified form in terms of the finite integral for all four binary modulations may be derived by utilizing the alternate expression of the complementary incomplete gamma function

$$\Gamma(\beta, z) = 2z^\beta \int_0^{\pi/2} \frac{\cos\theta}{\sin^{2\beta+1}\theta} \exp\left(-\frac{z}{\sin^2\theta}\right) d\theta \tag{5.5}$$

to express the CEP in (4.2) as [150]

$$P_{e,b} = \frac{(\alpha\gamma_b)^\beta}{\Gamma(\beta)} \int_0^{\pi/2} \frac{\cos\theta}{\sin^{2\beta+1}\theta} \exp\left(-\frac{\alpha\gamma_b}{\sin^2\theta}\right) d\theta \tag{5.6}$$

The average BEP may be obtained through averaging (5.6) over the PDF of γ_b given in (3.49)

$$
\begin{aligned}
P_{e,b} &= \int_0^\infty P_{e,b}(\gamma_b) f_\gamma(\gamma_b) d\gamma_b \\
&= \frac{\alpha^\beta}{\Gamma(\beta)} \frac{1+K}{\bar\gamma_b} \exp(-K) \int_0^{\pi/2} \frac{\cos\theta}{\sin^{2\beta+1}\theta} \\
&\quad \times \int_0^\infty \gamma_b^\beta \exp\left[-\gamma_b\left(\frac{\alpha}{\sin^2\theta} + \frac{1+K}{\bar\gamma_b}\right)\right] \\
&\quad \times I_0\left(2\sqrt{\frac{K(1+K)\gamma_b}{\bar\gamma_b}}\right) d\gamma_b d\theta
\end{aligned}
\tag{5.7}
$$

where the order of integrals has been interchanged to facilitate computation. Utilizing (B.11) the inner integral may be evaluated as

$$
\begin{aligned}
&\int_0^\infty \gamma_b^\beta \exp\left[-\gamma_b\left(\frac{\alpha}{\sin^2\theta} + \frac{1+K}{\bar\gamma_b}\right)\right] I_0\left(2\sqrt{\frac{K(1+K)\gamma_b}{\bar\gamma_b}}\right) d\gamma_b \\
&= \Gamma(\beta+1)\left[\frac{\bar\gamma_b \sin^2\theta}{\alpha\bar\gamma_b + (1+K)\sin^2\theta}\right]^{\beta+1} \\
&\quad \times {}_1F_1\left(\beta+1;1;\frac{K(1+K)\sin^2\theta}{\alpha\bar\gamma_b + (1+K)\sin^2\theta}\right)
\end{aligned}
\tag{5.8}
$$

Inserting the integration result in (5.7) we finally get

$$
\begin{aligned}
P_{e,b} &= \frac{\beta(1+K)}{2\alpha\bar\gamma_b}\exp(-K) \\
&\quad \int_0^{\pi/2} \sin(2\theta)\left[\frac{\alpha\bar\gamma_b}{\alpha\bar\gamma_b + (1+K)\sin^2\theta}\right]^{\beta+1} \\
&\quad \times {}_1F_1\left(\beta+1;1;\frac{K(1+K)\sin^2\theta}{\alpha\bar\gamma_b + (1+K)\sin^2\theta}\right)
\end{aligned}
\tag{5.9}
$$

Fig. 5.1 shows a plot of (5.9) for BPSK ($\alpha = 1, \beta = 1/2$) and DPSK ($\alpha = 1, \beta = 1$) modulations with three different fading parameters ($K = 0$, 6 dB, and 9 dB). It may be noted that the performance of BPSK and BFSK described here are special cases of the unified SEP expressions for coherent modulations provided in Section 5.5.

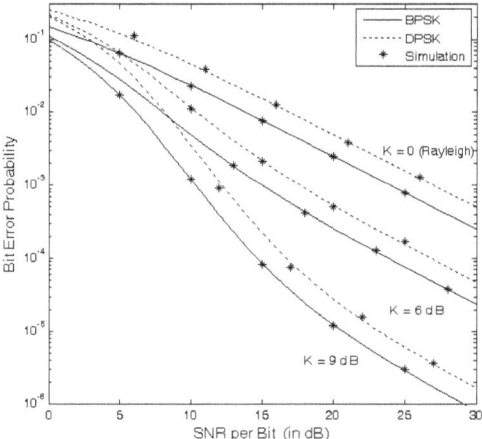

Figure 5.1: BEP of BPSK and DPSK in Rician fading.

5.2 Performance of $\pi/4$-DQPSK

Following the infinite series representation of modified Bessel functions [9, (9.6.12)], [8, (8.447.1)]

$$I_0(z) = \sum_{n=0}^{\infty} \frac{z^{2n}}{4^n (n!)^2} \tag{5.10}$$

the Rician fading PDF in (3.49) can be expressed as

$$
\begin{aligned}
f_\gamma(\gamma) &= \frac{(1+K)\exp(-K)}{\bar{\gamma}} \sum_{n=0}^{\infty} \left[\frac{K(1+K)}{\bar{\gamma}} \right]^n \\
&\quad \times \frac{\gamma^n}{(n!)^2} \exp\left[-\frac{\gamma(1+K)}{\bar{\gamma}} \right]
\end{aligned} \tag{5.11}
$$

Now the BEP of $\pi/4$-DQPSK over Rician fading channel may be found when the CEP given in (4.42) is averaged over (5.11). Noting that $\int_0^\infty f_\gamma(\gamma_b)d\gamma_b = 1$ and from the integration result provided in (A.4), the final average BEP for Rician fading becomes

$$
\begin{aligned}
P_{e,b} &= \frac{1}{2} \left[1 - \frac{\sqrt{2}\bar{\gamma}_b \exp(-K)}{1+K+2\bar{\gamma}_b} \sum_{n=0}^{\infty} \frac{K^n}{n!} \sum_{i=0}^{n} \left(\frac{1+K}{1+K+2\bar{\gamma}_b} \right)^i \right. \\
&\quad \left. \times {}_2F_1\left(\frac{i+1}{2}, \frac{i+2}{2}; 1; \frac{2\bar{\gamma}_b^2}{(1+K+2\bar{\gamma}_b)^2} \right) \right]
\end{aligned} \tag{5.12}
$$

139

In Fig. 5.2, BEP of differentially detected $\pi/4$-DQPSK operating over Rician fading channel is depicted for three different K values ($K = 0$, 6 dB and 9 dB) using (5.12). Specifically, $K = 0$ (-∞ dB) refers to the Rayleigh case given in (4.40).

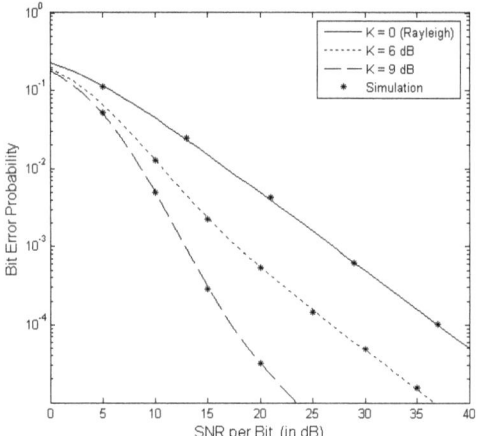

Figure 5.2: BEP of $\pi/4$-DQPSK in Rician fading.

Alternatively, from (4.50) and (3.49), using the method of finite integral

$$
\begin{aligned}
P_{e,b} &= \frac{(1 + K)\exp(-K)}{2\pi\bar{\gamma}_b} \int_{\gamma_b=0}^{\infty} \exp\left[-\frac{\gamma_b(1 + K)}{\bar{\gamma}_b}\right] \\
&\times I_0\left[2\sqrt{\frac{\gamma_b K(1 + K)}{\bar{\gamma}_b}}\right] \int_{\theta=0}^{\pi} \frac{\exp[-\gamma_b(2 - \sqrt{2}\cos\theta)]}{\sqrt{2} - \cos\theta}d\theta d\gamma_b
\end{aligned}
\tag{5.13}
$$

Interchanging order of integration

$$
\begin{aligned}
P_{e,b} &= \frac{(1 + K)\exp(-K)}{2\pi\bar{\gamma}_b} \int_{0}^{\pi} \frac{1}{\sqrt{2} - \cos\theta} \\
&\times \int_{0}^{\infty} \exp\left[-\gamma_b\left\{\frac{1 + K}{\bar{\gamma}_b} + (2 - \sqrt{2}\cos\theta)\right\}\right] \\
&\times I_0\left[2\sqrt{\frac{\gamma_b K(1 + K)}{\bar{\gamma}_b}}\right] d\gamma_b d\theta
\end{aligned}
\tag{5.14}
$$

The inner integral can be readily solved by using (B.12) resulting in

$$
\int_0^\infty \exp\left[-\gamma_b\left\{\frac{1+K}{\bar\gamma_b} + (2 - \sqrt{2}\cos\theta)\right\}\right] I_0\left[2\sqrt{\frac{\gamma_b K(1+K)}{\bar\gamma_b}}\right] d\gamma_b
$$
$$
= \frac{\bar\gamma_b}{1 + K + \bar\gamma_b(2 - \sqrt{2}\cos\theta)} \exp\left[\frac{K(1+K)}{1 + K + \bar\gamma_b(2 - \sqrt{2}\cos\theta)}\right] \tag{5.15}
$$

Hence, the final error expression in the form of finite integral

$$
P_{e,b} = \frac{1}{2\pi} \int_0^\pi \frac{G\exp[-K(1 - G)]}{\sqrt{2} - \cos\theta} d\theta \tag{5.16}
$$

where

$$
G = \frac{1+K}{1 + K + \bar\gamma_b(2 - \sqrt{2}\cos\theta)} \tag{5.17}
$$

as given in [73, (8.214)] [1] agreeing with [38, (6)]. It may be noted that the BEP of $\pi/4$-DQPSK over Rician fading channel was also mentioned in an earlier paper by Tjhung et al. (see (15) of [116]), where they have presented the BEP in the following form

$$
P_{e,b} = \frac{1}{\pi}\left[\frac{1}{2}\int_0^\pi \Psi(\theta)d\theta + \sum_{n=1}^\infty \left(\sqrt{2} - 1\right)^n \int_0^\pi \Psi(\theta)\cos(n\theta)d\theta\right] \tag{5.18}
$$

with

$$
\Psi(\theta) = \frac{1+K}{1 + K + \bar\gamma_b(2 - \sqrt{2}\cos\theta)} \exp\left[-\frac{K\bar\gamma_b(2 - \sqrt{2}\cos\theta)}{1 + K + \bar\gamma_b(2 - \sqrt{2}\cos\theta)}\right] \tag{5.19}
$$

5.2.1 BEP for MRC

For L-branch MRC diversity over IID Rician channels, the PDF of instantaneous SNR per symbol at combiner output γ_{MRC} is non-central chi-square distributed with $2L$ degrees of freedom and given by [66]

$$
f_{\gamma_{MRC}}(\gamma) = \frac{1+K}{\bar\gamma}\left[\frac{(1+K)\gamma}{LK\bar\gamma}\right]^{(L-1)/2} \exp\left[-LK - \frac{(1+K)\gamma}{\bar\gamma}\right]
$$
$$
\times I_{L-1}\left[2\sqrt{\frac{\gamma LK(1+K)}{\bar\gamma}}\right] \tag{5.20}
$$

where γ denotes instantaneous SNR per symbol per branch. From the IID assumption, we get that each branch has equal average branch SNR $\bar\gamma = E\{\gamma_k\}$ for $k \in \{1, 2, \cdots, L\}$,

---

[1]Actually the expression given in [73, (8.214)] is erroneous. The argument of the exponential term inside integration should not contain any negative sign.

average SNR at combiner output $\bar{\gamma}_{MRC} = \sum_{k=1}^{L} \bar{\gamma}_k = L\bar{\gamma}$, and that all diversity channels have an identical Rician factor, i.e. $K_k = K; \forall k$.

Like the single channel case, using the infinite series expansion of Bessel functions, namely

$$I_n(z) = \left(\frac{z}{2}\right)^n \sum_{m=0}^{\infty} \frac{z^{2m}}{4^m (m!)\Gamma(m+n+1)} \tag{5.21}$$

we can write (5.20) in the form

$$
\begin{aligned}
f_{\gamma_{MRC}}(\gamma) &= \exp(-LK) \sum_{n=0}^{\infty} \frac{(LK)^n}{(n!)\Gamma(n+L)} \left(\frac{1+K}{\bar{\gamma}}\right)^{n+L} \\
&\times \gamma^{n+L-1} \exp\left[-\frac{(1+K)\gamma}{\bar{\gamma}}\right]
\end{aligned}
\tag{5.22}
$$

Using (A.4), the average BEP of $\pi/4$-DQPSK with Lth order MRC diversity over Rician fading channel becomes

$$
\begin{aligned}
P_{e,b} &= \frac{1}{2}\left[1 - \frac{\sqrt{2}\bar{\gamma}_b \exp(-LK)}{1+K+2\bar{\gamma}_b}\right. \\
&\times \sum_{n=0}^{\infty} \frac{(LK)^n}{n!} \sum_{i=0}^{n+L-1} \left(\frac{1+K}{1+K+2\bar{\gamma}_b}\right)^i \\
&\left. \times \, {}_2F_1\left(\frac{i+1}{2}, \frac{i+2}{2}; 1; \frac{2\bar{\gamma}_b^2}{(1+K+2\bar{\gamma}_b)^2}\right)\right]
\end{aligned}
\tag{5.23}
$$

Fig. 5.3 shows the BEP for $\pi/4$-DQPSK over Rician fading channel with MRC diversity for various diversity orders ($L = 2, 3,$ and 4) when the fading parameter is kept fixed at $K = 6$ dB.

The corresponding finite integral solution may be obtained from (4.50) and (5.20) as

$$
\begin{aligned}
P_{e,b} &= \frac{LK \exp(-LK)}{2\pi} \left(\frac{1+K}{LK\bar{\gamma}_b}\right)^{(L+1)/2} \int_{\gamma_b=0}^{\infty} \gamma_b^{(L-1)/2} \\
&\times \exp\left[-\frac{\gamma_b(1+K)}{\bar{\gamma}_b}\right] I_{L-1}\left[2\sqrt{\frac{\gamma_b LK(1+K)}{\bar{\gamma}_b}}\right] \\
&\times \int_{\theta=0}^{\pi} \frac{\exp[-\gamma_b(2-\sqrt{2}\cos\theta)]}{\sqrt{2}-\cos\theta} d\theta d\gamma_b
\end{aligned}
\tag{5.24}
$$

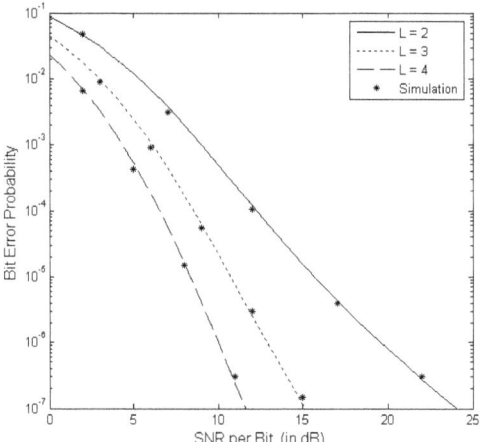

Figure 5.3: BEP of $\pi/4$-DQPSK in Rician ($K = 6$ dB) fading with MRC.

Interchanging order of integration we have

$$
\begin{aligned}
P_{e,b} &= \frac{LK\exp(-LK)}{2\pi}\left(\frac{1+K}{LK\bar{\gamma}_b}\right)^{(L+1)/2} \\
&\times \int_0^\pi \frac{1}{\sqrt{2}-\cos\theta}\int_0^\infty \gamma_b^{(L-1)/2}I_{L-1}\left[2\sqrt{\frac{\gamma_b LK(1+K)}{\bar{\gamma}_b}}\right] \\
&\times \exp\left[-\gamma_b\left\{\frac{1+K}{\bar{\gamma}_b}+(2-\sqrt{2}\cos\theta)\right\}\right]d\gamma_b d\theta
\end{aligned}
\tag{5.25}
$$

where the inner integral can be readily solved by using (B.11) resulting in

$$
\begin{aligned}
\mathcal{I} &= \left[\frac{\bar{\gamma}_b}{1+K+\bar{\gamma}_b(2-\sqrt{2}\cos\theta)}\right]^L\left[\frac{LK(1+K)}{\bar{\gamma}_b}\right]^{(L-1)/2} \\
&\times {}_1F_1\left(L;L;\frac{LK(1+K)}{1+K+\bar{\gamma}_b(2-\sqrt{2}\cos\theta)}\right)
\end{aligned}
\tag{5.26}
$$

After some manipulation and using [8, (9.215.1)], the final error expression is

$$
P_{e,b} = \frac{1}{2\pi}\int_0^\pi \frac{G^L\exp[-LK(1-G)]}{\sqrt{2}-\cos\theta}d\theta
\tag{5.27}
$$

where G is defined in (5.17).

5.3 Performance of Coherent MFSK

In case of a Rician channel the MGF $M_\gamma(s)$ is given by (3.51). Putting this value in (4.64) we can obtain the average error rate as

$$
\begin{aligned}
P_e &= \frac{M-1}{\pi} \int_0^{\pi/2} f_4(\theta) f_5(\theta) d\theta \\
&\quad - \frac{(M-1)(M-2)}{2\pi} \int_0^{\pi/4} f_6(\theta) f_7(\theta) d\theta \\
&\quad + \frac{(M-1)(M-2)(M-3)}{6\pi^2} \int_0^{\pi/6} f_6(\theta) f_2(\theta) f_7(\theta) d\theta \\
&\quad + \frac{(M-1)(M-2)(M-3)}{12\pi^2} \\
&\quad\quad \times \int_0^{\sin^{-1}\left(\frac{1}{\sqrt{3}}\right)} f_6(\theta)[\pi - f_2(\theta)] f_7(\theta) d\theta \\
&\quad - \frac{(M-1)(M-2)(M-3)(M-4)}{24\pi^2} \\
&\quad\quad \times \int_0^{\pi/6} f_6(\theta) f_2(\theta) f_7(\theta) d\theta
\end{aligned}
\tag{5.28}
$$

where

$$
f_4(\theta) = \frac{2(1+K)\sin^2\theta}{\bar{\gamma} + 2(1+K)\sin^2\theta}
\tag{5.29}
$$

$$
f_5(\theta) = \exp\left[-\frac{K\bar{\gamma}}{\bar{\gamma} + 2(1+K)\sin^2\theta}\right]
\tag{5.30}
$$

$$
f_6(\theta) = \frac{(1+K)\sin\theta\sqrt{1+\sin^2\theta}}{\bar{\gamma} + (1+K)(1+\sin^2\theta)}
\tag{5.31}
$$

$$
f_7(\theta) = \exp\left[-\frac{K\bar{\gamma}}{\bar{\gamma} + (1+K)(1+\sin^2\theta)}\right]
\tag{5.32}
$$

and $f_2(\theta)$ is defined in (2.112). For $K = 0$, as $f_6(\theta)$ becomes equal to $f_3(\theta)$ mentioned in (4.66) and $f_5(\theta) = f_7(\theta) = 1$, (5.28) reduces to the SEP expression for Rayleigh channel in (4.65) as expected. Direct evaluation from the PDF is also possible by using (B.12) and performing integration of all five terms.

In Fig. 5.4, SEP of coherent MFSK over Rician (Nakagami-n) channel given by (5.28) is depicted for various values of modulation order M (=2, 4, and 10). Fig. 5.4 also demonstrates the effect of K (= 3 dB and 9 dB) on SEP.

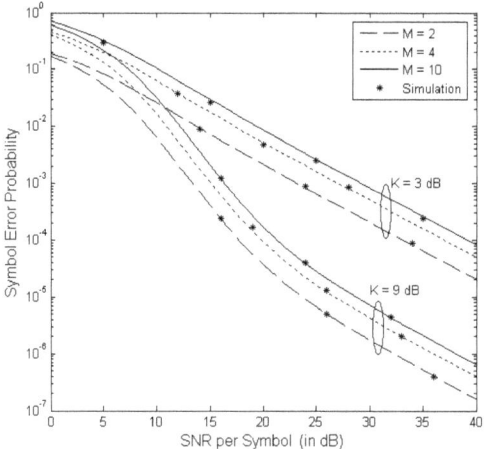

Figure 5.4: SEP of coherent MFSK in Rician fading.

5.3.1 SEP for MRC

Substituting the MGF for Rician channel in (4.68) we get the SEP of coherent MFSK for Rician channel with MRC diversity as

$$
\begin{aligned}
P_e &= \frac{M-1}{\pi} \int_0^{\pi/2} g_3(\theta)d\theta \\
&\quad - \frac{(M-1)(M-2)}{2\pi} \int_0^{\pi/4} g_4(\theta)d\theta \\
&\quad + \frac{(M-1)(M-2)(M-3)}{6\pi^2} \int_0^{\pi/6} g_4(\theta)f_2(\theta)d\theta \\
&\quad + \frac{(M-1)(M-2)(M-3)}{12\pi^2} \int_0^{\sin^{-1}\left(\frac{1}{\sqrt{3}}\right)} g_4(\theta)[\pi - f_2(\theta)]d\theta \\
&\quad - \frac{(M-1)(M-2)(M-3)(M-4)}{24\pi^2} \int_0^{\pi/6} g_4(\theta)f_2(\theta)d\theta
\end{aligned}
\tag{5.33}
$$

where

$$
\begin{aligned}
g_3(\theta) &= \left[\frac{2(1+K)\sin^2\theta}{\bar{\gamma} + 2(1+K)\sin^2\theta}\right]^L \\
&\quad \times \exp\left[-\frac{LK\bar{\gamma}}{\bar{\gamma} + 2(1+K)\sin^2\theta}\right]
\end{aligned}
\tag{5.34}
$$

145

$$g_4(\theta) = \frac{\sin\theta}{\sqrt{1+\sin^2\theta}} \left[\frac{(1+K)(1+\sin^2\theta)}{\bar{\gamma}+(1+K)(1+\sin^2\theta)} \right]^L$$
$$\times \exp\left[-\frac{LK\bar{\gamma}}{\bar{\gamma}+(1+K)(1+\sin^2\theta)} \right] \tag{5.35}$$

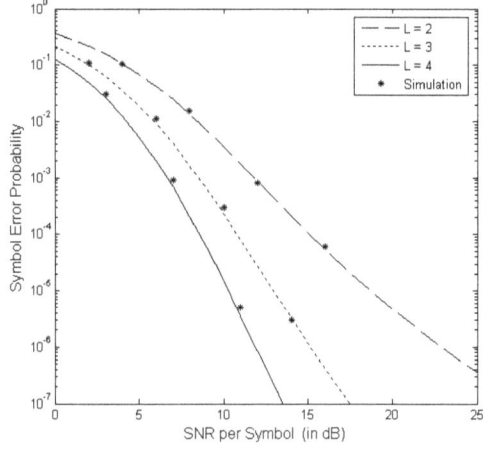

Figure 5.5: SEP of coherent MFSK ($M = 8$) in Rician ($K = 6$ dB) fading with MRC.

Fig. 5.5 shows the coherent MFSK error rates over Rician fading channel with MRC diversity. The pattern of SEP variation with modulation order M or fading parameter K is well studied in non-diversity cases (see Fig. 5.4) and hence are not repeated for the MRC case. Accordingly, keeping the constellation size ($M = 8$) and fading parameter ($K = 6$ dB) fixed, SEP values are depicted for various diversity orders ($L = 2$, 3, and 4).

5.4   Performance of Non-coherent MFSK

The average SEP of non-coherent MFSK over single Rician fading channel can be derived by substituting the Rician PDF given in (3.49) and the CEP for non-coherent MFSK

$$P_e(\gamma) = \sum_{j=1}^{M-1} \frac{(-1)^{j+1}}{j+1} \binom{M-1}{j} \exp\left(-\frac{j\gamma}{j+1} \right) \tag{5.36}$$

from (2.128) in the generic equation $\int_0^\infty P_e(\gamma)f_\gamma(\gamma)d\gamma$ as

$$
\begin{aligned}
P_e &= \frac{1+K}{\bar{\gamma}}\exp(-K)\sum_{j=1}^{M-1}\frac{(-1)^{j+1}}{j+1}\binom{M-1}{j}\\
&\quad \times \int_0^\infty \exp\left[-\gamma\left(\frac{j}{j+1}+\frac{1+K}{\bar{\gamma}}\right)\right]\\
&\quad \times I_0\left(2\sqrt{\frac{K(1+K)\gamma}{\bar{\gamma}}}\right)d\gamma
\end{aligned}
\tag{5.37}
$$

The integral in (5.37) may be solved using (B.12) to yield

$$
\begin{aligned}
P_e &= \sum_{j=1}^{M-1}\frac{(-1)^{j+1}(1+K)}{j\bar{\gamma}+(j+1)(1+K)}\binom{M-1}{j}\\
&\quad \times \exp\left[-\frac{jK\bar{\gamma}}{j\bar{\gamma}+(j+1)(1+K)}\right]
\end{aligned}
\tag{5.38}
$$

For $K = 0$, i.e. for Rayleigh fading, (5.38) reduces to

$$
P_e = \sum_{j=1}^{M-1}\frac{(-1)^{j+1}}{1+j(1+\bar{\gamma})}\binom{M-1}{j}
\tag{5.39}
$$

It is easy to verify that for $M = 2$ (ncBFSK), (5.38) and (5.39) boils down to (5.4) with $\alpha = 1/2$ and [27, (6.159)] respectively.

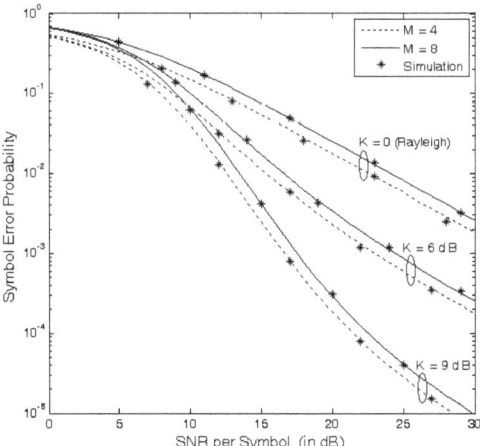

Figure 5.6: SEP of ncMFSK in Rician fading.

In Fig. 5.6 SEP performance of ncMFSK over Rician fading channel is plotted for different fading parameters and for different constellation sizes. From the plots we can conclude that at a given SNR $\bar{\gamma}$ and M, as K value is increased, SEP decreases due to the presence of strong LOS component. On the other hand, at a particular SNR and keeping the fading parameter constant, as M increases SEP becomes more as a higher constellation requires more SNR to attain a particular value of symbol error.

5.4.1 SEP for SC

For selection combiner operating over Rician channel, the SEP cannot be found in closed-form through the simple PDF method as described above. Instead, we make use of the CDF method

$$P_e = -\int_0^\infty P_e'(\gamma)F_{\gamma_{SC}}(\gamma)d\gamma \tag{5.40}$$

and further substitute γ with $\tan\theta$ following the approach by Haghani and Beaulieu [151, 152]

$$P_e = -\int_0^{\pi/2} P_e'(\tan\theta)F_{\gamma_{SC}}(\tan\theta)\sec^2\theta d\theta \tag{5.41}$$

to convert the integral into a finite range one so that subsequent numerical integration is possible. Now inserting

$$P_e'(\gamma) = \sum_{j=1}^{M-1} \frac{j(-1)^{j+2}}{(j+1)^2}\binom{M-1}{j}\exp\left(-\frac{j\gamma}{j+1}\right) \tag{5.42}$$

which was obtained by differentiating (5.36) with respect to γ, and

$$F_{\gamma_{SC}}(\gamma) = [F_\gamma(\gamma)]^L$$

$$= \left[1 - Q\left(\sqrt{2K}, \sqrt{\frac{2(1+K)}{\bar{\gamma}}}\tan\theta\right)\right]^L \tag{5.43}$$

in (5.41) we obtain finally

$$P_e = \sum_{j=1}^{M-1} \frac{j(-1)^{j+1}}{(j+1)^2} \binom{M-1}{j}$$
$$\times \int_0^{\pi/2} \exp\left(-\frac{j\tan\theta}{j+1}\right) \tag{5.44}$$
$$\times \left[1 - Q\left(\sqrt{2K}, \sqrt{\frac{2(1+K)}{\bar{\gamma}}}\tan\theta\right)\right]^L \sec^2\theta d\theta$$

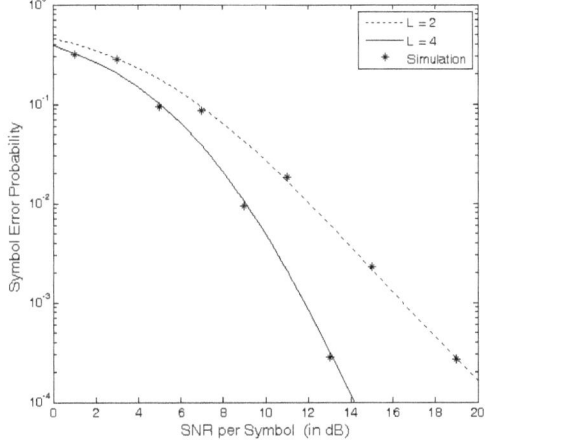

Figure 5.7: SEP of ncMFSK ($M = 4$) in Rician ($K - 3$ dB) fading with SC.

In Fig. 5.7, SEP for nc4FSK in Rician fading channel with SC are shown where the advantage of using a higher order diversity is clearly visible. When 4 antennas are used, the chances of receiving a correct symbol increases compared to the dual diversity case. Effect of varying K and M are considered in the previous plot and hence are not repeated here.

5.5 Unified Approach for SEP Analysis

The conditional SEP $P_e(\gamma_b)$, or BEP $P_{e,b}(\gamma_b)$, for several coherent modulation techniques such as MQAM, QPSK, its offset version OQPSK, minimum shift keying (MSK), and DBPSK involves the term $\text{erfc}^2(\sqrt{k\gamma_b})$. The presence of $\text{erfc}^2(\cdot)$ in conditional SEP makes the integral $\int_0^\infty P_e(\gamma_b)f_\gamma(\gamma_b)d\gamma_b$ difficult to evaluate, and it is a common practice to drop

Modulation type	Value of parameters		
	a	b	c
Square MQAM	$\frac{2(\sqrt{M}-1)}{\sqrt{M}}$	$\frac{2\sqrt{M}-1}{M}-1$	$\frac{3}{2}\frac{\log_2 M}{M-1}$
QPSK/ OQPSK/ MSK	1	$-\frac{1}{4}$	1
DBPSK	1	$-\frac{1}{2}$	1
BPSK	$\frac{1}{2}$	0	1
BFSK	$\frac{1}{2}$	0	$\frac{1}{2}$

Table 5.1: CEP parameter values for the unified expression.

the term for simplification. However when SNR (γ_b) is low, the term has noticeable effect on average SEP [65]. In case of MQAM the approximation become quite crude when the constellation size (M) becomes larger.

A general expression of conditional SEP for coherent modulation schemes involving $\text{erfc}^2(\cdot)$ may be written in the following form

$$P_e(\gamma_b) = a\ \text{erfc}(\sqrt{c\gamma_b}) + b\ \text{erfc}^2(\sqrt{c\gamma_b}) \tag{5.45}$$

where we assume that complete CSI (fading amplitude, phase, and delay) is available at the receiver. The parameter values a, b, and c for different modulations can be found from Table 5.1. Using the generalized $P_e(\gamma_b)$ expression given in (5.45) accurate average SEP is calculated in the later subsections.

5.5.1 Generalized SEP

In this subsection, we derive a unified exact SEP expression for coherent modulation schemes operating in Rician fading environment, based on the direct averaging of the conditional SEP over the PDF of the received signal envelope. Exploiting the relationship [9, (7.1.21)]

$$\text{erfc}(z) = 1 - \text{erf}(z) = 1 - \frac{2z}{\sqrt{\pi}}\exp(-z^2)\,{}_1F_1\left(1;\frac{3}{2};z^2\right) \tag{5.46}$$

the conditional SEP in (5.45) can be written in terms of confluent hypergeometric function $_1F_1(\cdot;\cdot;\cdot)$ as

$$
\begin{aligned}
P_e(\gamma_b) &= a + b - 2(a + 2b)\sqrt{\frac{c\gamma_b}{\pi}}\exp(-c\gamma_b){}_1F_1\left(1;\frac{3}{2};c\gamma_b\right) \\
&\quad + \frac{4bc\gamma_b}{\pi}\exp(-2c\gamma_b)\left[{}_1F_1\left(1;\frac{3}{2};c\gamma_b\right)\right]^2
\end{aligned}
\tag{5.47}
$$

and the average SEP can be found by substituting the Rician PDF in series form in (5.11), and (5.47), in $\int_0^\infty P_e(\gamma_b)f_\gamma(\gamma_b)d\gamma_b$ to obtain

$$
P_e = a + b - 2(a + 2b)\sqrt{\frac{c}{\pi}}\mathcal{I}_1 + \frac{4bc}{\pi}\mathcal{I}_2
\tag{5.48}
$$

where

$$
\begin{aligned}
\mathcal{I}_1 &= \frac{1+K}{\bar{\gamma}_b}\exp(-K)\sum_{n=0}^{\infty}\frac{K^n(1+K)^n}{\bar{\gamma}_b^n(n!)^2} \\
&\quad \times \int_0^\infty \gamma_b^{n+1/2}\exp\left[-\gamma_b\left(c + \frac{1+K}{\bar{\gamma}_b}\right)\right] \\
&\quad \times {}_1F_1\left(1;\frac{3}{2};c\gamma_b\right)d\gamma_b
\end{aligned}
\tag{5.49}
$$

and

$$
\begin{aligned}
\mathcal{I}_2 &= \frac{1+K}{\bar{\gamma}_b}\exp(-K)\sum_{n=0}^{\infty}\frac{K^n(1+K)^n}{\bar{\gamma}_b^n(n!)^2} \\
&\quad \times \int_0^\infty \gamma_h^{n+1}\exp\left[-\gamma_b\left(2c + \frac{1+K}{\bar{\gamma}_b}\right)\right] \\
&\quad \times \left[{}_1F_1\left(1;\frac{3}{2};c\gamma_b\right)\right]^2 d\gamma_b
\end{aligned}
\tag{5.50}
$$

It may be noted that the infinite series representation of $I_0(\cdot)$ in (5.10) has been used to obtain the above forms.

The integral $\mathcal{I}_1$ can be readily evaluated utilizing [8, (7.621.4)]

$$
\begin{aligned}
\mathcal{I}_1 &= \frac{1+K}{1+K+c\bar{\gamma}_b}\exp(-K)\sqrt{\frac{\bar{\gamma}_b}{1+K+c\bar{\gamma}_b}} \\
&\quad \times \sum_{n=0}^{\infty}\frac{\Gamma(n+3/2)}{(n!)^2}\left[\frac{K(1+K)}{1+K+c\bar{\gamma}_b}\right]^n \\
&\quad \times {}_2F_1\left(1,n+\frac{3}{2};\frac{3}{2};\frac{c\bar{\gamma}_b}{1+K+c\bar{\gamma}_b}\right)
\end{aligned}
\tag{5.51}
$$

Further, with the help of transformation formula [9, (15.3.4)] of the Gaussian hypergeo-metric function, $\mathcal{I}_1$ can be further reduced to

$$
\begin{aligned}
\mathcal{I}_1 &= \sqrt{\frac{\bar{\gamma}_b}{1 + K + c\bar{\gamma}_b}} \exp(-K) \sum_{n=0}^{\infty} \frac{\Gamma(n + 3/2)}{(n!)^2} \\
&\times \left[\frac{K(1 + K)}{1 + K + c\bar{\gamma}_b}\right]^n {}_2F_1\left(1, -n; \frac{3}{2}; -\frac{c\bar{\gamma}_b}{1 + K}\right)
\end{aligned}
\tag{5.52}
$$

This reduction not only simplifies the final form but also improves the stability and convergence of the series solutions to a great extent.

The evaluation of $\mathcal{I}_2$ is not so straightforward as the integral consists of square of ${}_1F_1(\cdot; \cdot; \cdot)$. The integration may be performed using [67, (10)] and [67, (13)] first in terms of Appell's hypergeometric function $F_2(\cdot; \cdot, \cdot; \cdot, \cdot; \cdot, \cdot)$

$$
\begin{aligned}
\mathcal{I}_2 &= \frac{1 + K}{\bar{\gamma}_b} \exp(-K) \left(\frac{\bar{\gamma}_b}{1 + K + 2c\bar{\gamma}_b}\right)^2 \\
&\times \sum_{n=0}^{\infty} \frac{\Gamma(n + 2)}{(n!)^2} \left[\frac{K(1 + K)}{1 + K + 2c\bar{\gamma}_b}\right]^n \\
&\times F_2\left(n + 2; 1, 1; \frac{3}{2}, \frac{3}{2}; \frac{c\bar{\gamma}_b}{1 + K + 2c\bar{\gamma}_b}, \frac{c\bar{\gamma}_b}{1 + K + 2c\bar{\gamma}_b}\right)
\end{aligned}
\tag{5.53}
$$

and then in terms of Gauss hypergeometric function as

$$
\begin{aligned}
\mathcal{I}_2 &= \frac{\bar{\gamma}_b \exp(-K)}{1 + K + 2c\bar{\gamma}_b} \sum_{n=0}^{\infty} \frac{K^n}{n!} \sum_{j=0}^{n} \left(\frac{1 + K}{1 + K + 2c\bar{\gamma}_b}\right)^j \\
&\times {}_2F_1\left(1, j + 1; \frac{3}{2}; \frac{c\bar{\gamma}_b}{1 + K + 2c\bar{\gamma}_b}\right)
\end{aligned}
\tag{5.54}
$$

Finally inserting the expressions of (5.52) and (5.54) in (5.48) one may get the complete unified SEP expression

$$
\begin{aligned}
P_e &= a + b - \sqrt{\frac{c\bar{\gamma}_b}{\pi}} \exp(-K) \sum_{n=0}^{\infty} \frac{K^n}{n!} \\
&\times \left\{ \frac{a + 2b}{\sqrt{1 + K + c\bar{\gamma}_b}} \frac{\Gamma(n + 3/2)}{n!} \right. \\
&\times \left(\frac{1 + K}{1 + K + c\bar{\gamma}_b}\right)^n {}_2F_1\left(1, -n; \frac{3}{2}; -\frac{c\bar{\gamma}_b}{1 + K}\right) \\
&- \frac{2b}{1 + K + 2c\bar{\gamma}_b} \sqrt{\frac{c\bar{\gamma}_b}{\pi}} \sum_{j=0}^{n} \left(\frac{1 + K}{1 + K + 2c\bar{\gamma}_b}\right)^j \\
&\left. \times {}_2F_1\left(1, j + 1; \frac{3}{2}; \frac{c\bar{\gamma}_b}{1 + K + 2c\bar{\gamma}_b}\right)\right\}
\end{aligned}
\tag{5.55}
$$

The generalized Rician channel SEP expression with $K = 0$ becomes the SEP for the Rayleigh fading case. By setting $K = 0$ in (5.52) and (5.54) we have

$$\mathcal{I}_1 = \frac{1}{2}\sqrt{\frac{\pi\bar{\gamma}_b}{1 + c\bar{\gamma}_b}} \tag{5.56}$$

$$\mathcal{I}_2 = \frac{\bar{\gamma}_b}{1 + 2c\bar{\gamma}_b} \, {}_2F_1\left(1, 1; \frac{3}{2}; \frac{c\bar{\gamma}_b}{1 + 2c\bar{\gamma}_b}\right) \tag{5.57}$$

Using [9, (15.3.4)] and [9, (15.1.5)], and after some algebra, one may attain an equivalent form of the hypergeometric function in (5.57) as given by

$$_2F_1\left(1, 1; \frac{3}{2}; \frac{c\bar{\gamma}_b}{1 + 2c\bar{\gamma}_b}\right) = \frac{1 + 2c\bar{\gamma}_b}{1 + c\bar{\gamma}_b}\sqrt{\frac{1 + c\bar{\gamma}_b}{c\bar{\gamma}_b}}\tan^{-1}\left(\sqrt{\frac{c\bar{\gamma}_b}{1 + c\bar{\gamma}_b}}\right) \tag{5.58}$$

which yields the final expression of $\mathcal{I}_2$ in the form

$$\mathcal{I}_2 = \sqrt{\frac{\bar{\gamma}_b}{c(1 + c\bar{\gamma}_b)}}\tan^{-1}\left(\sqrt{\frac{c\bar{\gamma}_b}{1 + c\bar{\gamma}_b}}\right) \tag{5.59}$$

Thus the SEP for Rayleigh fading is

$$\begin{aligned}
P_e &= (a + b) - (a + 2b)\sqrt{\frac{c\bar{\gamma}_b}{1 + c\bar{\gamma}_b}} \\
&\quad + \frac{4b}{\pi}\sqrt{\frac{c\bar{\gamma}_b}{1 + c\bar{\gamma}_b}}\tan^{-1}\left(\sqrt{\frac{c\bar{\gamma}_b}{1 + c\bar{\gamma}_b}}\right)
\end{aligned} \tag{5.60}$$

The validity of the result can be shown by setting the parameter values for BPSK ($a = 1/2$, $b = 0$, and $c = 1$) in (5.60)

$$P_{e,b} = \frac{1}{2}\left[1 - \sqrt{\frac{\bar{\gamma}_b}{1 + \bar{\gamma}_b}}\right] \tag{5.61}$$

which is a well established result [2, (14-3-7)].

5.5.2   Generalized SEP with MRC

Mimicking the approach in Section 5.5.1, the average SEP of different coherent modulations can be found in an unified manner by substituting (5.22) and (5.47) in $\int_0^\infty P_e(\gamma_b)f_{\gamma_{MRC}}(\gamma_b)d\gamma_b$. Although the form of final SEP expression remains same as in (5.48), the values of the integrals are different as described below

$$
\begin{aligned}
\mathcal{I}_1 &= \exp(-LK) \sum_{n=0}^{\infty} \frac{(LK)^n}{n!\Gamma(n+L)} \left(\frac{1+K}{\bar{\gamma}_b}\right)^{m+L} \\
&\quad \times \int_0^{\infty} \gamma_b^{n+L-1/2} \exp\left[-\gamma_b\left(c+\frac{1+K}{\bar{\gamma}_b}\right)\right] \\
&\quad \times {}_1F_1\left(1;\frac{3}{2};c\gamma_b\right) d\gamma
\end{aligned}
\tag{5.62}
$$

$$
\begin{aligned}
\mathcal{I}_2 &= \exp(-LK) \sum_{n=0}^{\infty} \frac{(LK)^n}{n!\Gamma(n+L)} \left(\frac{1+K}{\bar{\gamma}_b}\right)^{m+L} \\
&\quad \times \int_0^{\infty} \gamma_b^{n+L} \exp\left[-\gamma_b\left(2c+\frac{1+K}{\bar{\gamma}_b}\right)\right] \\
&\quad \times \left[{}_1F_1\left(1;\frac{3}{2};c\gamma_b\right)\right]^2 d\gamma
\end{aligned}
\tag{5.63}
$$

Using [8, (7.621.4)] and [9, (15.3.4)], the integral $\mathcal{I}_1$ can be calculated as

$$
\begin{aligned}
\mathcal{I}_1 &= \exp(-LK)\sqrt{\frac{\bar{\gamma}_b}{1+K+c\bar{\gamma}_b}} \sum_{n=0}^{\infty} \frac{(LK)^n}{n!} \frac{\Gamma(n+L+1/2)}{\Gamma(n+L)} \\
&\quad \times \left[\frac{K(1+K)}{1+K+c\bar{\gamma}_b}\right]^{n+L-1} {}_2F_1\left(1,1-n-L;\frac{3}{2};-\frac{c\bar{\gamma}_b}{1+K}\right)
\end{aligned}
\tag{5.64}
$$

which reduces to (5.52) for $L = 1$. Following Falujah and Prabhu's approach [67], the integral in $\mathcal{I}_2$ can be solved in series form as

$$
\begin{aligned}
\mathcal{I}_2 &= \exp(-LK) \sum_{n=0}^{\infty} \frac{(LK)^n \Gamma(n+L+1)}{n!\Gamma(n+L)} \\
&\quad \times \left(\frac{1+K}{\bar{\gamma}_b}\right)^{m+L} \left(\frac{\bar{\gamma}_b}{1+K+2c\bar{\gamma}_b}\right)^{n+L+1} \\
&\quad \times F_2\left(n+L+1;1,1;\frac{3}{2},\frac{3}{2};\frac{c\bar{\gamma}_b}{1+K+2c\bar{\gamma}_b},\frac{c\bar{\gamma}_b}{1+K+2c\bar{\gamma}_b}\right)
\end{aligned}
\tag{5.65}
$$

Finally, exploiting the properties of Appell's function [67], $\mathcal{I}_2$ can be simplified as

$$
\begin{aligned}
\mathcal{I}_2 &= \frac{\bar{\gamma}_b \exp(-LK)}{1+K+2c\bar{\gamma}_b} \sum_{n=0}^{\infty} \frac{(LK)^n}{n!} \sum_{j=0}^{n+L-1} \left(\frac{1+K}{1+K+2c\bar{\gamma}_b}\right)^j \\
&\quad \times {}_2F_1\left(1,j+1;\frac{3}{2};\frac{c\bar{\gamma}_b}{1+K+2c\bar{\gamma}_b}\right)
\end{aligned}
\tag{5.66}
$$

As expected, by putting $L = 1$ in (5.66) one can easily get back to (5.54). Combining (5.64), (5.66), and (5.48), the complete unified SEP expression for coherent modulations over Rician fading channel with MRC is

$$
\begin{aligned}
P_e &= a + b - \sqrt{\frac{c\bar{\gamma}_b}{\pi}}\exp(-LK)\sum_{n=0}^{\infty}\frac{(LK)^n}{n!} \\
&\quad \times \left\{ \frac{a+2b}{\sqrt{1+K+c\bar{\gamma}_b}}\frac{\Gamma(n+L+1/2)}{\Gamma(n+L)}\left(\frac{1+K}{1+K+c\bar{\gamma}_b}\right)^{n+L-1}\right. \\
&\qquad\qquad \times {}_2F_1\left(1,1-n-L;\frac{3}{2};-\frac{c\bar{\gamma}_b}{1+K}\right) \\
&\quad -\frac{2b}{1+K+2c\bar{\gamma}_b}\sqrt{\frac{c\bar{\gamma}_b}{\pi}}\sum_{j=0}^{n+L-1}\left(\frac{1+K}{1+K+2c\bar{\gamma}_b}\right)^{j} \\
&\qquad\qquad \left.\times {}_2F_1\left(1,j+1;\frac{3}{2};\frac{c\bar{\gamma}_b}{1+K+2c\bar{\gamma}_b}\right)\right\}
\end{aligned}
\tag{5.67}
$$

By setting $K = 0$ in (5.67), we can obtain the error rate expressions of the said coherent modulations with MRC, operating in a Rayleigh fading channel. After some manipulations and noting that

$$
\frac{\Gamma\left(n+\frac{1}{2}\right)}{\Gamma\left(\frac{1}{2}\right)} = \frac{\Gamma(2n+1)}{2^{2n}\Gamma(n+1)}
\tag{5.68}
$$

the SEP for multichannel Rayleigh fading scenario may be written as

$$
\begin{aligned}
P_e &= a + b - \sqrt{\frac{c\bar{\gamma}_b}{1+c\bar{\gamma}_b}}\binom{2L}{L}\frac{L(a+2b)}{2[4(1+c\bar{\gamma}_b)]^{L-1}} \\
&\qquad \times {}_2F_1\left(1,1-L;\frac{3}{2};-c\bar{\gamma}_b\right) \\
&\quad +\frac{4bc\bar{\gamma}_b}{\pi}\sum_{j=0}^{L-1}\frac{1}{(1+2c\bar{\gamma}_b)^{j+1}}\,{}_2F_1\left(1,j+1;\frac{3}{2};\frac{c\bar{\gamma}_b}{1+2c\bar{\gamma}_b}\right)
\end{aligned}
\tag{5.69}
$$

A special case of which (for MQAM) was provided by Kim et al. [63]. As a double check, it can be easily verified that (5.69) reduces to (5.60) for $L = 1$.

5.5.3 Performance of MQAM

From the generalized SEP expressions formulated in Section 5.5.1 and in Section 5.5.2, we may easily obtain the SEP expression for square MQAM, by setting the proper parameter values as given in Table 5.1. Before presenting the graphical results, however, let us review the related literature in order to show the improvement attained through the results presented in this book, both in terms of accuracy and tractability.

Following the MGF and CHF approach respectively, Patterh et al. [153] and Yang et al. [154] evaluated SEP for MQAM over Rician channels with MRC. However, their

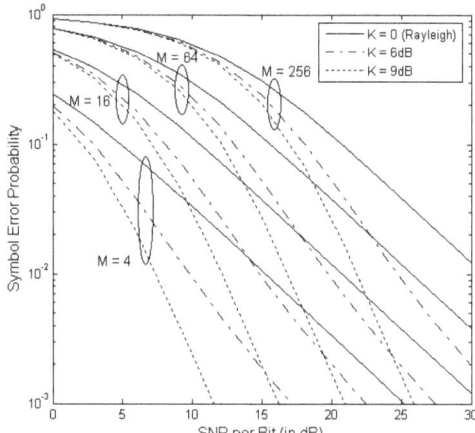

Figure 5.8: SEP of square MQAM in Rician fading.

end expressions involved several versions of a single integral, which have to be computed by numerical methods. Ali and Alkhudairi [155], on the other hand, presented BEP for MQAM in Rician fading with MRC, following an approximation by Goldsmith and Chua [156]. The approximation accuracy (within 1 dB) was claimed for AWGN channels only and that too within a small SNR range (0-30 dB). Obviously there exists no guarantee that the same degree of precision would be followed in a Rician channel as well. Some researchers [157, 158, 159] investigated the problem by decomposing a square MQAM constellation into in-phase and quadrature PAM signal sets of constellation size of $\sqrt{M}$ and thereby avoiding the square of the complementary error function $\text{erfc}^2(\cdot)$ in the integrand. Nevertheless, the solutions provide nothing more than upper bounds and not applicable to modulation schemes like DBPSK, which cannot be decomposed into components in such a way.

It therefore appears highly desirable to have a simple, general, closed-form treatment of the square MQAM error rate. In this connection, we would like to mention that the expressions developed in this book covers a general class of modulations including square MQAM, reduces to the well known Rayleigh fading results, and provides dual expressions presented by Seo et al. [66] and Lu et al. [64]. The expressions available in Seo's paper [66] were relatively complex as it consisted of both confluent and Gaussian hypergeometric functions. Also, their solutions lead to numerical instability as well as divergent series in some cases.

From the error expression in (5.55), the average SEP versus SNR per bit $\bar{\gamma}_b$ for square MQAM is plotted in Fig. 5.8 for various combinations of the constellation size $M(= 4,$

156

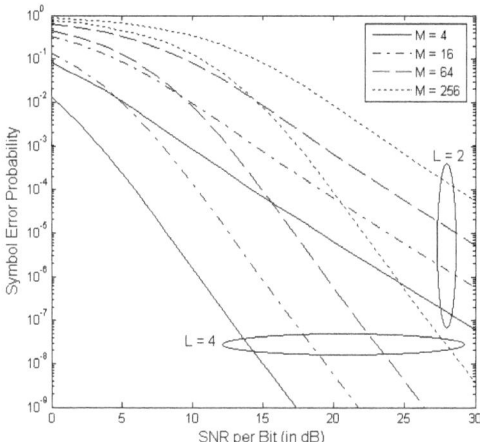

Figure 5.9: SEP of square MQAM in Rician ($K = 3$ dB) fading with MRC.

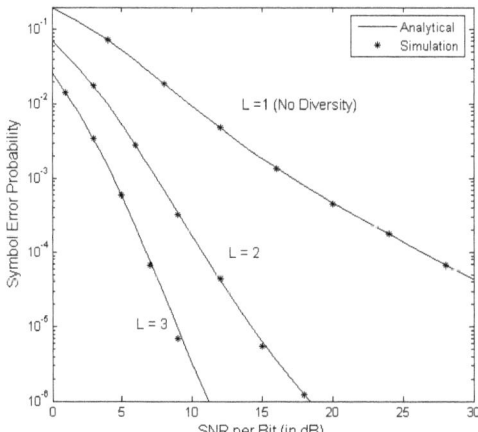

Figure 5.10: Analytical and Simulated SEP of square 4QAM in Rician ($K = 6$ dB) fading.

16, 64 and 256) and fading parameter $K(= 0$, 6 dB and 9 dB) values. It is clear from the figure that for a fixed value of M, as the fading parameter value K increases, lesser transmission power is needed in order to achieve the same SEP. The most severe fading ($K = 0$) denotes Rayleigh fading scenario which arises when the specular component vanishes and reception is non-LOS only. At the other extreme, as $K \to \infty$, the SEP approaches AWGN channel error performance. For a fixed value of K, as M increases, to keep the SEP constant, there is a demand for higher SNR which is quite consistent with the theory of M-ary modulation.

In Fig. 5.9, SEP of square MQAM operating over Rician fading channels ($K = 3$ dB) with MRC diversity ($L = 2$ and 4) is depicted for various orders of modulation ($M = 4$, 16, 64 and 256) using (5.67). Inspecting the SEP curves one finds that when SNR, order of modulation M, and fading parameter K are kept constant, the SEP lowers as number of diversity branches (L) increases.

In Fig. 5.10, SEP performance in a Rician channel ($K = 6$ dB) with MRC diversity for 4-ary QAM is depicted. Second order ($L = 2$), third order ($L = 3$) along with the no diversity case ($L = 1$) have been considered. Since the curves obtained from analytical standpoint almost coincide with their simulation counterpart, this observation validates the accuracy of our derivations.

5.5.4 Performance of other Coherent Modulations

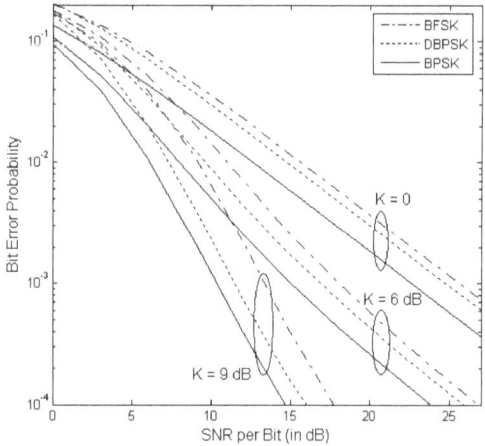

Figure 5.11: BEP of DBPSK, BPSK and BFSK in Rician fading.

From Table 5.1 one can identify that the expression of SEP for QPSK (or OQPSK /MSK) becomes identical with 4-QAM when the parameter value $M = 4$ is substituted in the SEP expression of MQAM. Hence QPSK plots are not shown separately. BEP for DBPSK, BPSK and BFSK for different K values are shown in Fig. 5.11. The results for BPSK and BFSK match exactly with available results [147] and have been included for

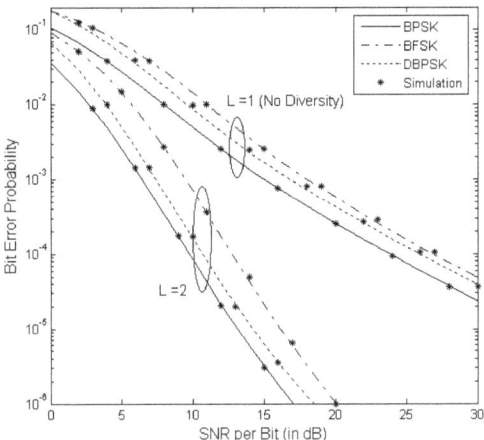

Figure 5.12: Analytical and Simulated BEP of DBPSK, BPSK and BFSK in Rician ($K = 6$ dB) fading.

comparison purpose only. Like the MQAM case, in Fig. 5.11 a similar effect of increasing K value can be observed for every modulation scheme.

In Fig. 5.12 the BEP values for DBPSK, BPSK, and BFSK with MRC combining are shown for a fixed fading parameter, $K = 6$ dB. An exact match between the analytical and simulated BEP is observed for both single ($L = 1$) and multichannel ($L = 2$) reception.

Chapter Summary

In this chapter, we derived BEP and SEP expressions for different modulation schemes operating over Rician fading channel. Simpler series solutions for BEP of $\pi/4$-DQPSK with MRC, are provided. The series, although infinite, converges fast and by considering different number of terms of the series, one may trade off accuracy with complexity as well as speed of computation. In addition, SEP of both coherent and non-coherent MFSK is found in presence of suitable diversity combining (MRC and SC respectively). Compact integral form SEP expressions are given for both the cases. A unified approach for evaluating error performance of several coherent modulation schemes with MRC diversity is proposed next. The analytical framework includes BPSK, BFSK, QPSK, OQPSK, MSK, and square MQAM as a special cases, thus dispensing with the need for SEP calculation for individual modulations separately.

Chapter 6

Diversity Combining in Nakagami Fading

Nakagami's m-distribution serves as a versatile statistical model for characterizing time-varying multipath channels. By varying the fading parameter m, which ranges from $1/2$ to ∞, the model can accurately fit experimental data obtained for urban, semi-urban, and rural land mobile propagations. Due to its wide applicability and relatively simple PDF expression for instantaneous channel gain, Nakagami model has attracted attention for many researchers in the past few decades. For example, BEP of binary modulations with SC and MRC are discussed in [160] and [161], respectively. Calculations for MPSK without diversity was provided by Aalo and Pattaramalai [162], and later extended to cover multichannel SC [163] and MRC reception [164]. MDPSK is a differentially coherent modulation technique, and switched diversity is a viable option for the said modulation. Fedele, in his 1998 paper [165], derived SEP of MDPSK with switched diversity over Nakagami fading. On the other hand, the performance of MDPSK with MRC were investigated separately by Chai and Tjhung [166], and by Ekanayake et al. [167]. Switched diversity is also applicable to non-coherent MFSK modulation, and was addressed by Yang et al. [168]. In this connection, another paper by Radaydeh and Matalgah [169] is worth mentioning, where the performance of ncMFSK with non-coherent EGC was derived. Finally, for MQAM modulation, we would also like to cite some of the existing papers by Annamalai et al. [65], Falujah and Prabhu [67], Patterh et al. [170], and Yoon et al. [171]. All these papers dealt with diversity reception with MQAM modulation over Nakagami channel under different conditions.

From the foregoing discussion on Nakagami channel model, it is evident that there is a little scope for any new contribution regarding error probability analysis of digital modulations with diversity combining. The earlier contributions are not included in this book except in Section 6.1, where we conduct a brief review of BEP calculation for binary modulations in a unified manner. In the next section, Section 6.2, BEP performance of

$\pi/4$-DQPSK with MRC and SC diversity are presented. A series solution for the MRC case is provided, while we were able to find a closed-form expression for the SEP with SC diversity. The next section, Section 6.3, is devoted for some improved approximations for SEP of coherent MFSK with MRC diversity.

6.1 Performance of Binary Modulations

As shown in Section 2.3.1, CEP for all the four binary modulations may be expressed using the following unified expression [7, 172]

$$P_{e,b}(\gamma_b) = \frac{\Gamma(\beta, \alpha\gamma_b)}{2\Gamma(\beta)} \tag{6.1}$$

where the notations have their usual meanings. The set of relations $\Gamma(1/2, z^2) = \sqrt{\pi}\mathrm{erfc}(z)$, and $\Gamma(1, z) = \exp(-z)$ accounts for the validity of the unified expression in connection with coherent $(\beta = 1/2)$ and non-coherent $(\beta = 1)$ modulation types.

The unified expression for the probability of error may be found by averaging this CEP over the Nakagami PDF mentioned in (3.54)

$$\begin{aligned} P_{e,b} &= \int_0^\infty P_{e,b}(\gamma_b) f_\gamma(\gamma_b) d\gamma_b \\ &= \frac{1}{2\Gamma(\beta)\Gamma(m)} \left(\frac{m}{\bar{\gamma}_b}\right)^m \int_0^\infty \Gamma(\beta, \alpha\gamma_b)\gamma_b^{m-1} \exp\left(-\frac{m\gamma_b}{\bar{\gamma}_b}\right) d\gamma_b \end{aligned} \tag{6.2}$$

which may be solved using [8, (6.455.1)] to yield

$$\begin{aligned} P_{e,b} &= \frac{1}{2\Gamma(\beta)\Gamma(m)} \left(\frac{m}{\bar{\gamma}_b}\right)^m \frac{\alpha^\beta \Gamma(m+\beta)}{m\left(\alpha + \frac{m}{\bar{\gamma}_b}\right)^{m+\beta}} \\ &\quad \times {}_2F_1\left(1, m+\beta; m+1; \frac{m}{m+\alpha\bar{\gamma}_b}\right) \end{aligned} \tag{6.3}$$

In the above expression, all the parameters in the hypergeometric function depend on a single parameter m. Thus further simplification may be possible which can be accomplished by expressing hypergeometric function with incomplete beta function $B_z(\cdot, \cdot)$ using [9, (26.5.23)]

$$ {}_2F_1(a, 1-b; a+1; z) = \frac{a}{z^a} I n_z(a, b) B(a, b) = \frac{a}{z^a} B_z(a, b) \frac{\Gamma(\beta, \alpha\gamma_b)}{2\Gamma(\beta)} \tag{6.4}$$

to obtain

$$P_{e,b} = \frac{\Gamma(m+\beta)}{2\Gamma(m)\Gamma(\beta)}B_{m/(m+\alpha\bar{\gamma}_b)}(m,\beta) = \frac{B_{m/(m+\alpha\bar{\gamma}_b)}(m,\beta)}{2B(m,\beta)}$$
$$= \frac{1}{2}In_{m/(m+\alpha\bar{\gamma}_b)}(m,\beta) \qquad (6.5)$$

where $In_z(a,b) = B_z(a,b)/B(a,b)$ [9, (6.6.2)] denotes the ratio of Beta functions, and the well known relation between Beta function and Gamma function namely, $\Gamma(a)\Gamma(b)/\Gamma(a+b) = B(a,b)$ [9, (6.2.2)] has been used for the derivation.

Putting $n = a$ in the series formula [9, (6.6.4)]

$$In_z(a, n-a+1) = \sum_{j=a}^{n}\binom{n}{j}z^j(1-z)^{n-j} \qquad (6.6)$$

we get

$$In_z(a,1) = z^a \qquad (6.7)$$

Now, using this property, we can express $P_{e,b}$ for binary non-coherent type modulations as

$$P_{e,b} = \frac{1}{2}\left(\frac{m}{m+\alpha\bar{\gamma}_b}\right)^m \qquad (6.8)$$

which reduces to

$$P_{e,b} = \frac{1}{2}\left(\frac{m}{m+\bar{\gamma}_b}\right)^m \qquad (6.9)$$

for DPSK, and

$$P_{e,b} = \frac{1}{2}\left(\frac{2m}{2m+\bar{\gamma}_b}\right)^m \qquad (6.10)$$

for ncBFSK. Alternatively (6.9) and (6.10) can be derived in an independent manner using the integral representation of Gamma function by (with η_b replaced by γ_b) averaging (2.47) and (2.48) over the Nakagami PDF. It may be noted that for $m = 1$, standard results for Rayleigh fading channel [27, (6.158)] and [27, (6.159)] are obtained.

Actually, the unified expression in (6.5) is helpful for evaluating the performances of binary coherent modulations in closed form. Putting $\beta = 1/2$ in the unified expression

$$
\begin{aligned}
P_{e,b} &= \frac{In_{m/(m+\alpha\bar{\gamma}_b)}\left(m, \frac{1}{2}\right)}{2} = \frac{B_{m/(m+\alpha\bar{\gamma}_b)}\left(m, \frac{1}{2}\right)}{2B\left(m, \frac{1}{2}\right)} \\
&= \frac{\Gamma\left(m + \frac{1}{2}\right)}{2\Gamma(m)\Gamma\left(\frac{1}{2}\right)} B_{m/(m+\alpha\bar{\gamma}_b)}\left(m, \frac{1}{2}\right)
\end{aligned}
\tag{6.11}
$$

Further using

$$
\frac{\Gamma\left(n + \frac{1}{2}\right)}{\Gamma\left(\frac{1}{2}\right)} = \frac{\Gamma(2n+1)}{2^{2n}\Gamma(n+1)}
\tag{6.12}
$$

which may be easily derived from the recurrence formula $\Gamma(z+1) = z\Gamma(z)$, and [7, (A.3)]

$$
B_z(a,b) = \frac{z^a}{a}(1-z)^b \, {}_2F_1(1, a+b; a+1; z)
\tag{6.13}
$$

we can write

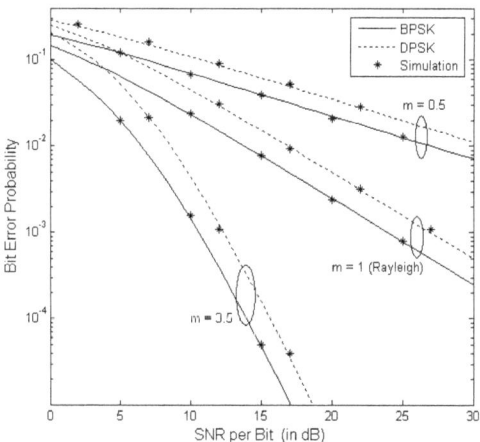

Figure 6.1: BEP of BPSK and DPSK in Nakagami fading.

$$
\begin{aligned}
P_{e,b} &= \frac{\Gamma(2m)}{\Gamma(m+1)\Gamma(m)}\left(\frac{m/4}{m+\alpha\bar{\gamma}_b}\right)^m \left(\frac{\alpha\bar{\gamma}_b}{m+\alpha\bar{\gamma}_b}\right)^{1/2} \\
&\quad \times {}_2F_1\left(1, m + \frac{1}{2}; m+1; \frac{m}{m+\alpha\bar{\gamma}_b}\right)
\end{aligned}
\tag{6.14}
$$

163

which reduces to

$$P_{e,b} = \frac{\Gamma(2m)}{\Gamma(m+1)\Gamma(m)} \left[\frac{m}{4(m+\bar{\gamma}_b)}\right]^m \left(\frac{\bar{\gamma}_b}{m+\bar{\gamma}_b}\right)^{1/2}$$
$$\times {}_2F_1\left(1, m+\frac{1}{2}; m+1; \frac{m}{m+\bar{\gamma}_b}\right)$$

(6.15)

for BPSK, and

$$P_{e,b} = \frac{\Gamma(2m)}{\Gamma(m+1)\Gamma(m)} \left(\frac{m}{4m+2\bar{\gamma}_b}\right)^m \left(\frac{\bar{\gamma}_b}{2m+\bar{\gamma}_b}\right)^{1/2}$$
$$\times {}_2F_1\left(1, m+\frac{1}{2}; m+1; \frac{2m}{2m+\bar{\gamma}_b}\right)$$

(6.16)

for BFSK. The probability of error for BPSK with $m = 1$

$$P_{e,b} = \frac{1}{4(1+\bar{\gamma}_b)} \sqrt{\frac{\bar{\gamma}_b}{1+\bar{\gamma}_b}} \, {}_2F_1\left(1, \frac{3}{2}; 2; \frac{1}{1+\bar{\gamma}_b}\right)$$

(6.17)

Using the relation [9, (15.1.14)]

$$_2F_1\left(a, a+\frac{1}{2}; 2a; z\right) = 2^{2a-1}(1-z)^{-1/2}\left[1+(1-z)^{1/2}\right]^{1-2a}$$

(6.18)

from (6.17)

$$P_{e,b} = \frac{1}{2}\left[1 - \sqrt{\frac{\bar{\gamma}_b}{1+\bar{\gamma}_b}}\right]$$

(6.19)

which is a standard result for Rayleigh fading channel [27, (6.156)]. Similarly, it can be also shown that the formula for BFSK boils down to the familiar form [27, (6.157)]. The BEP for BPSK and DPSK as a function of average SNR per bit is displayed in Fig. 6.1 for three different fading parameter values.

6.2 Performance of $\pi/4$-DQPSK

When m = integer

Averaging (2.93) over the Nakagami PDF while using the integration result given in (A.4), the BEP expression can be found in closed-form

$$P_{e,b} = \frac{1}{2}\left[1 - \frac{\sqrt{2}\bar{\gamma}_b}{m+2\bar{\gamma}_b}\sum_{i=0}^{m-1}\left(\frac{m}{m+2\bar{\gamma}_b}\right)^i \times {}_2F_1\left(\frac{i+1}{2},\frac{i+2}{2};1;\frac{2\bar{\gamma}_b^2}{(m+2\bar{\gamma}_b)^2}\right)\right]$$

(6.20)

Using the identity [8, (9.121.1)] it can be shown that for $m = 1$, i.e., for the Rayleigh fading case, (6.20) reduces to the standard form in (4.40).

When $m \neq$ integer

When m is not an integer, the error expression can be obtained in the form of an infinite series using the CEP in series form given in (4.33)

$$P_{e,b} = \mathcal{I}_1 - \mathcal{I}_2$$

(6.21)

where

$$\mathcal{I}_1 = \frac{1}{\Gamma(m)}\left(\frac{m}{\bar{\gamma}_b}\right)^m\sum_{j=0}^{\infty}\frac{1}{(\sqrt{2}+1)^j} \times \int_0^{\infty}\gamma_b^{m-1}\exp\left[-\gamma_b\left(2+\frac{m}{\bar{\gamma}_b}\right)\right]I_j\left(\sqrt{2}\gamma_b\right)d\gamma_b$$

(6.22)

and

$$\mathcal{I}_2 = \frac{1}{2\Gamma(m)}\left(\frac{m}{\bar{\gamma}_b}\right)^m\int_0^{\infty}\exp\left[-\gamma_b\left(2+\frac{m}{\bar{\gamma}_b}\right)\right]I_0\left(\sqrt{2}\gamma_b\right)d\gamma_b$$

(6.23)

From (B.3) the first integral

$$\mathcal{I}_1 = \frac{1}{\Gamma(m)}\left(\frac{m}{m+2\bar{\gamma}_b}\right)^m\sum_{j=0}^{\infty}\left[\frac{\bar{\gamma}_b}{(2+\sqrt{2})(m+2\bar{\gamma}_b)}\right]^j \times \frac{\Gamma(m+j)}{\Gamma(j+1)}{}_2F_1\left(\frac{m+j+1}{2},\frac{m+j}{2};j+1;\frac{2\bar{\gamma}_b^2}{(m+2\bar{\gamma}_b)^2}\right)$$

(6.24)

Likewise the second integral can be evaluated using (B.4)

$$\mathcal{I}_2 = \frac{1}{2\Gamma(m)}\left(\frac{m}{\bar{\gamma}_b}\right)^m\frac{\bar{\gamma}_b}{\sqrt{m^2+4m\bar{\gamma}_b+2\bar{\gamma}_b^2}}$$

(6.25)

Hence the final expression of average BEP

$$
\begin{aligned}
P_{e,b} &= \frac{m^m}{\Gamma(m)} \left[\frac{1}{(m + 2\bar{\gamma}_b)^m} \right. \\
&\qquad \times \sum_{j=0}^{\infty} \left\{ \frac{\bar{\gamma}_b}{(2 + \sqrt{2})(m + 2\bar{\gamma}_b)} \right\}^j \frac{\Gamma(m + j)}{\Gamma(j + 1)} \\
&\qquad \times {}_2F_1\left(\frac{m + j + 1}{2}, \frac{m + j}{2}; j + 1; \frac{2\bar{\gamma}_b^2}{(m + 2\bar{\gamma}_b)^2} \right) \\
&\qquad \left. - \frac{1}{2\gamma_b^m} \frac{\bar{\gamma}_b}{\sqrt{m^2 + 4m\bar{\gamma}_b + 2\bar{\gamma}_b^2}} \right]
\end{aligned}
\tag{6.26}
$$

Fig. 6.2 plots the analytical results from (6.26) giving BEP over Nakagami fading channel. Three different fading parameter (m) values ($m = 0.8$, 1.5, and 3.0) are considered.

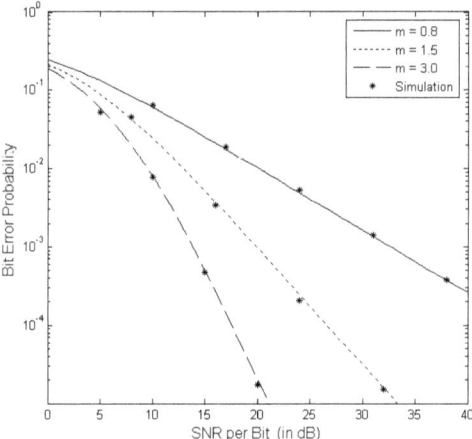

Figure 6.2: BEP of $\pi/4$-DQPSK in Nakagami fading.

On the other hand, averaging (4.50) over the Nakagami PDF in (3.54) results in

$$
\begin{aligned}
P_{e,b} &= \frac{1}{2\pi\Gamma(m)} \left(\frac{m}{\bar{\gamma}_b} \right)^m \int_0^{\pi} \frac{1}{\sqrt{2} - \cos\theta} \\
&\qquad \times \int_0^{\infty} \gamma_b^{m-1} \exp\left[-\gamma_b \frac{m + \bar{\gamma}_b(2 - \sqrt{2}\cos\theta)}{\bar{\gamma}_b} \right] d\gamma_b d\theta
\end{aligned}
\tag{6.27}
$$

where a change of integration has been performed to express $P_{e,b}$ in the form given above.

The integration with respect to γ_b can be identified as a simple gamma function

$$\int_0^\infty \gamma_b^{m-1} \exp\left[-\gamma_b \frac{m + \bar{\gamma}_b(2 - \sqrt{2}\cos\theta)}{\bar{\gamma}_b}\right] d\gamma_b$$
$$= \Gamma(m)\left[\frac{\bar{\gamma}_b}{m + \bar{\gamma}_b(2 - \sqrt{2}\cos\theta)}\right]^m \tag{6.28}$$

which, when inserted in (6.27), gives

$$P_{e,b} = \frac{1}{2\pi}\int_0^\pi \frac{G^m}{\sqrt{2} - \cos\theta} d\theta \tag{6.29}$$

where

$$G = \frac{m}{m + \bar{\gamma}_b(2 - \sqrt{2}\cos\theta)} \tag{6.30}$$

as demonstrated by Simon and Alouini [73, (8.218)], and Tellambura and Bhargava [38, (7)]

$$P_{e,b} = \frac{\left(\frac{m}{m+2\bar{\gamma}_b}\right)^m}{2\pi}\int_0^\pi \frac{1}{\left(\sqrt{2} - \cos\theta\right)\left(1 - \frac{\sqrt{2\bar{\gamma}_b}}{m+2\bar{\gamma}_b}\cos\theta\right)^m} d\theta \tag{6.31}$$

and agreeing with the expression provided by Tanda [117, (11)],

$$P_{e,b} = \frac{1}{\pi}\left[\frac{1}{2}\int_0^\pi \Psi(\theta)d\theta + \sum_{n=1}^\infty \left(\sqrt{2} - 1\right)^n \int_0^\pi \Psi(\theta)\cos(n\theta)d\theta\right] \tag{6.32}$$

where

$$\Psi(\theta) = \left[1 + \frac{2\bar{\gamma}_b}{m}\left(1 - \frac{\cos\theta}{\sqrt{2}}\right)\right]^{-m} \tag{6.33}$$

6.2.1 BEP for MRC

In Nakagami fading channel, when L-order MRC diversity is employed at the receiver to mitigate the effect of multipath fading, the PDF of the combiner output SNR is given by [162]

$$f_\gamma(\gamma) = \left(\frac{m}{\bar{\gamma}}\right)^{Lm} \frac{\gamma^{Lm-1}}{\Gamma(Lm)} \exp\left(-\frac{m\gamma}{\bar{\gamma}}\right) \qquad ; \gamma \geq 0 \tag{6.34}$$

When m = integer

When m is an integer, the BEP expression can be found in closed-form using (A.4)

$$P_{e,b} = \frac{1}{2}\left[1 - \frac{\sqrt{2}\bar{\gamma}_b}{m + 2\bar{\gamma}_b} \sum_{i=0}^{Lm-1} \left(\frac{m}{m + 2\bar{\gamma}_b}\right)^i \right.$$
$$\left. \times {}_2F_1\left(\frac{i+1}{2}, \frac{i+2}{2}; 1; \frac{2\bar{\gamma}_b^2}{(m + 2\bar{\gamma}_b)^2}\right)\right] \tag{6.35}$$

which reduces to (6.20) for $L = 1$.

When $m \neq$ integer

When m is not an integer, the error expression can be obtained in the form of an infinite series like the single channel case

$$P_{e,b} = \mathcal{I}_1 - \mathcal{I}_2 \tag{6.36}$$

where

$$\mathcal{I}_1 = \frac{1}{\Gamma(Lm)}\left(\frac{m}{\bar{\gamma}_b}\right)^{Lm} \sum_{j=0}^{\infty} \frac{1}{(\sqrt{2}+1)^j}$$
$$\times \int_0^\infty \gamma_b^{Lm-1} \exp\left[-\gamma_b\left(2 + \frac{m}{\bar{\gamma}_b}\right)\right] I_j\left(\sqrt{2}\gamma_b\right) d\gamma_b \tag{6.37}$$

and

$$\mathcal{I}_2 = \frac{1}{2\Gamma(Lm)}\left(\frac{m}{\bar{\gamma}_b}\right)^{Lm} \int_0^\infty \exp\left[-\gamma_b\left(2 + \frac{m}{\bar{\gamma}_b}\right)\right] I_0\left(\sqrt{2}\gamma_b\right) d\gamma_b \tag{6.38}$$

From (B.3) the first integral

$$\mathcal{I}_1 = \frac{1}{\Gamma(Lm)}\left(\frac{m}{m + 2\bar{\gamma}_b}\right)^{Lm}$$
$$\times \sum_{j=0}^{\infty}\left[\frac{\bar{\gamma}_b}{(2 + \sqrt{2})(m + 2\bar{\gamma}_b)}\right]^j \frac{\Gamma(Lm + j)}{\Gamma(j+1)}$$
$$\times {}_2F_1\left(\frac{Lm+j+1}{2}, \frac{Lm+j}{2}; j+1; \frac{2\bar{\gamma}_b^2}{(m + 2\bar{\gamma}_b)^2}\right) \tag{6.39}$$

Likewise the second integral can be evaluated using (B.4)

$$\mathcal{I}_2 = \frac{1}{2\Gamma(Lm)}\left(\frac{m}{\bar{\gamma}_b}\right)^m \frac{\bar{\gamma}_b}{\sqrt{m^2 + 4m\bar{\gamma}_b + 2\bar{\gamma}_b^2}} \tag{6.40}$$

Inserting (6.39) and (6.40) in (6.36), the final expression of average BEP

$$
\begin{aligned}
P_{e,b} = {} & \frac{m^{Lm}}{\Gamma(Lm)} \Bigg[\frac{1}{(m + 2\bar{\gamma}_b)^m} \\
& \times \sum_{j=0}^{\infty} \left\{ \frac{\bar{\gamma}_b}{(2 + \sqrt{2})(m + 2\bar{\gamma}_b)} \right\}^j \frac{\Gamma(Lm + j)}{\Gamma(j + 1)} \\
& \times {}_2F_1\left(\frac{Lm + j + 1}{2}, \frac{Lm + j}{2}; j + 1; \frac{2\bar{\gamma}_b^2}{(m + 2\bar{\gamma}_b)^2} \right) \\
& - \frac{1}{2\gamma_b^{Lm}} \frac{\bar{\gamma}_b}{\sqrt{m^2 + 4m\bar{\gamma}_b + 2\bar{\gamma}_b^2}} \Bigg]
\end{aligned}
\tag{6.41}
$$

Fig. 6.3 shows the BEP for $\pi/4$-DQPSK over Nakagami fading channel with MRC diversity. Keeping the fading parameter ($m = 1.5$) fixed, BEP values are depicted for various diversity orders ($L = 2$, 3, and 4).

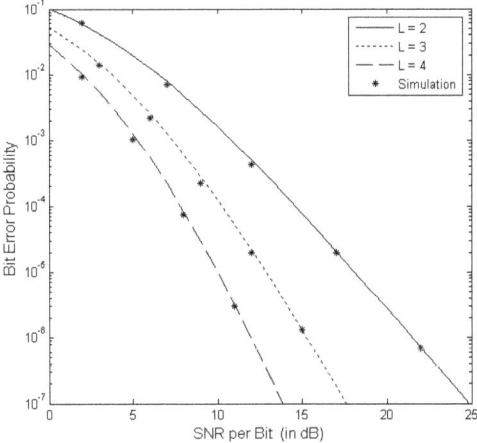

Figure 6.3: BEP of $\pi/4$-DQPSK in Nakagami ($m = 1.5$) fading with MRC.

Alternatively, averaging (4.50) over the PDF given in (6.34)

$$
\begin{aligned}
P_{e,b} = {} & \frac{1}{2\pi\Gamma(Lm)} \left(\frac{m}{\bar{\gamma}_b} \right)^m \int_0^\pi \frac{1}{\sqrt{2} - \cos\theta} \\
& \times \int_0^\infty \gamma_b^{Lm-1} \exp\left[-\gamma_b \frac{m + \bar{\gamma}_b(2 - \sqrt{2}\cos\theta)}{\bar{\gamma}_b} \right] d\gamma_b d\theta
\end{aligned}
\tag{6.42}
$$

After integrating the inner integral, the final expression becomes

$$P_{e,b} = \frac{1}{2\pi} \int_0^\pi \frac{G^{Lm}}{\sqrt{2} - \cos\theta} d\theta \qquad (6.43)$$

where G is as defined in (6.30).

6.2.2 BEP for SC

In this subsection, following Fedele's idea for binary case [173], closed-form BEP expressions are derived for $\pi/4$-DQPSK operating over IID slow nonselective Nakagami fading channels employing SC diversity for integer values of m. The error expression of BEP is in the form of a finite series that includes Gaussian hypergeometric function along with some other elementary functions, and can be calculated easily. The major contribution is that we have addressed the BEP, and not SEP, of differentially detected $\pi/4$-DQPSK with general L branch SC, and found a true closed-form expression for it.

Related Work

Earlier works on calculation of BEP over Nakagami fading channel with SC includes the following references [174, 173, 175, 176] and [160, 177, 163, 178, 179]. Blanco [174] and Okui [175] were among the first to address the PDF of dual diversity ($L = 2$) SC in Nakagami fading channel for integer values of fading parameter m. The corresponding CHF was derived in the later paper while the former one dealt with the error performance of ncBFSK. An extension to the general L branch case for ncMFSK with arbitrary m values was attempted by Sannegowda and Aalo [176]. Similar derivations were also reported later on by Ugweje [180] and Annavajjala et al. [181]. In all these papers, the error probability was derived in terms of Lauricella's multiple hypergeometric function $F_A(\cdot; \cdots; \cdots; \cdots)$ [8, (9.19)], which reduces to the well known Gaussian hypergeometric function only for $L = 2$ [180] or for $M = 2$ [176]. Further, Lauricella's function is not readily available in most mathematical softwares. In [180] and [181], the computations were performed using Gauss-Laguerre quadrature (GLQ) method, a numerical approximation technique, which becomes too tedious for $L > 2$ [180]. In another paper Ugweje and Aalo [177] expressed the combiner output SNR PDF in terms of Laguerre polynomials $L_n^\alpha(\cdot)$ [8, (8.970)] and calculated the error rates for binary signalling in correlated Nakagami fading channel with SC. The BER expression involved infinite series with poor convergence rate and again calculations become too cumbersome for $L > 2$. The approach of Baik et al. [163] was different, but resulted in similar kind of expressions with L fold summation. Actually the first paper that describes the approach followed in this book was presented by Fedele [173], and later by Lo and Lam [178, 179]. However, it may be noted that none of these papers dealt with the BEP of $\pi/4$-DQPSK.

While the PDF of SC combiner output in Nakagami fading channel is thoroughly described in open literature, there are not many papers dealing with error probability evaluation of DQPSK. In [182], Fedele described the SEP of MDPSK (DQPSK is a special case of MDPSK for $M = 4$) in Nakagami fading channel without employing diversity, and provided more general results for dual diversity SC [183] and L order diversity [184] later on. Although there were no restrictions on the value of m, but the SEP results were not in closed-form; SEP expressions included a finite range integral which needs to be evaluated numerically. Lo and Lam [185], on the other hand, came up with an approximation to avoid infinite sum or numerical integration.

In this regard, some recent research works [186, 187, 188, 189, 190, 191] are also worth mentioning. In particular Reig et al. [189] obtained error rates in correlated Nakagami fading with dual diversity SC in terms of Appell's hypergeometric function of second kind $F_2(\cdot; \cdot, \cdot; \cdot, \cdot; \cdot, \cdot)$ [8, (9.180.2)], whereas Kong and Milstein [191] presented an approximated SEP expression. The evaluation of BEP of differentially detected $\pi/4$-DQPSK is mentioned only in [186], [187] and [188], out of which Choo and Tjhung's result [186] includes infinite series, while in the rest two papers [187] and [188] BEP value is given with either single or multifold integral. Thus, a general framework giving closed-form BEP results for differentially detected $\pi/4$-DQPSK in Nakagami fading channel with SC is lacking till date.

PDF of Selection Combiner Output SNR

In diversity assisted fading channel communication, for receivers employing L-order SC, the CDF of the combiner output SNR is given by

$$F_{\gamma_{SC}}(\gamma) - [F_\gamma(\gamma)]^L \tag{6.44}$$

and the corresponding PDF can be obtained by differentiating the CDF as

$$f_{\gamma_{SC}}(\gamma) = \frac{\partial F_{\gamma_{SC}}(\gamma)}{\partial \gamma} = L \left[F_\gamma(\gamma) \right]^{L-1} f_\gamma(\gamma) \tag{6.45}$$

where $f_\gamma(\gamma)$ and $F_\gamma(\gamma)$ are the PDF and CDF for single fading channel respectively. For Nakagami channel, inserting the PDF and CDF from (3.54) and (3.55), the expression for the PDF of combiner output SNR becomes [192]

$$f_{\gamma_{SC}}(\gamma) = \frac{L}{[\Gamma(m)]^L} \left(\frac{m}{\bar{\gamma}} \right)^m \gamma^{m-1} \exp\left(-\frac{m\gamma}{\bar{\gamma}} \right) \left[\gamma\left(m, \frac{m\gamma}{\bar{\gamma}} \right) \right]^{L-1} \tag{6.46}$$

where $\gamma(\cdot, \cdot)$ is the incomplete gamma function.

For integer values of the fading parameter $m \ (= 1, 2, \ldots)$, the PDF can be reduced to

$$
\begin{aligned}
f_{\gamma_{SC}}(\gamma) \ = \ & \frac{L}{\Gamma(m)} \left(\frac{m}{\bar{\gamma}} \right)^m \gamma^{m-1} \exp\left(-\frac{m\gamma}{\bar{\gamma}} \right) \\
& \times \left[1 - \exp\left(-\frac{m\gamma}{\bar{\gamma}} \right) \sum_{j=0}^{m-1} \frac{(m\gamma/\bar{\gamma})^j}{j!} \right]^{L-1}
\end{aligned}
\tag{6.47}
$$

where we have used the following relation [8, (8.352.1)]

$$
\gamma(n+1, x) = n! \left[1 - \exp(-x) \sum_{j=0}^{n} \left(\frac{x^j}{j!} \right) \right]
\tag{6.48}
$$

to obtain the result. As demonstrated by Fedele [173], further simplification of (6.47) is possible when the second term is expressed in a series form using binomial theorem

$$
\begin{aligned}
& \left[1 - \exp\left(-\frac{m\gamma}{\bar{\gamma}} \right) \sum_{j=0}^{m-1} \frac{(m\gamma/\bar{\gamma})^j}{j!} \right]^{L-1} \\
& = \sum_{k=0}^{L-1} (-1)^k \binom{L-1}{k} \exp\left(-\frac{mk\gamma}{\bar{\gamma}} \right) \left[\sum_{j=0}^{m-1} \frac{(m\gamma/\bar{\gamma})^j}{j!} \right]^k
\end{aligned}
\tag{6.49}
$$

Also utilizing the relation [8, (0.314)]

$$
\left(\sum_{k=0}^{\infty} a_k x^k \right)^n = \sum_{k=0}^{\infty} c_k x^k
\tag{6.50}
$$

where $c_0 = (a_0)^n$ and

$$
c_m = \frac{1}{m a_0} \sum_{k=1}^{m} (kn - m + k) a_k c_{m-k} \qquad ; m \geq 1
\tag{6.51}
$$

the last term of (6.49) may be written as

$$
\left[\sum_{j=0}^{m-1} \frac{(m\gamma/\bar{\gamma})^j}{j!} \right]^k = \sum_{j=0}^{k(m-1)} b_j \left(\frac{m\gamma}{\bar{\gamma}} \right)^j
\tag{6.52}
$$

where $a_j = 1/j!$ and b_j is calculated using the recurrence relation

$$
b_j = \frac{1}{j} \sum_{i=1}^{\min(j, m-1)} \left[\frac{i(k+1) - j}{i!} \right] b_{j-1}
\tag{6.53}
$$

The initial values of these coefficients are $a_0 = 1$ and $b_0 = (a_0)^k = 1$, while the last term

172

$b_{k(m-1)}$ is obtained when the last term of the right hand sum in (6.52), i.e. $(m\gamma/\bar{\gamma})^{m-1}/(m-1)!$ is multiplied k times giving a value $b_{k(m-1)} = [1/(m-1)!]^k$. Finally inserting (6.52) in (6.49) and substituting (6.49) in (6.47), the expression for PDF can be written as

$$f_{\gamma_{sc}}(\gamma) = \frac{L}{\Gamma(m)} \sum_{k=0}^{L-1} (-1)^k \binom{L-1}{k}$$
$$\times \sum_{j=0}^{k(m-1)} b_j \left(\frac{m}{\bar{\gamma}}\right)^{j+m} \gamma^{j+m-1} \exp\left[-\frac{m\gamma(1+k)}{\bar{\gamma}}\right] \tag{6.54}$$

BEP with SC over Nakagami Fading Channel

The BEP of $\pi/4$-DQPSK over Nakagami fading channel with L order SC may be found when the CEP given in (4.42) is averaged over (6.54). Noting that $\int_0^\infty f_\gamma(\gamma_b)d\gamma_b = 1$, and from the integration result provided in (A.4), the final average BEP becomes

$$P_{e,b} = \frac{1}{2}\left[1 - \frac{L}{\Gamma(m)} \sum_{k=0}^{L-1} \frac{\sqrt{2}\bar{\gamma}_b(-1)^k}{m(1+k)+2\bar{\gamma}_b} \binom{L-1}{k}\right.$$
$$\times \sum_{j=0}^{k(m-1)} b_j \frac{(j+m-1)!}{(1+k)^{j+m}} \sum_{i=0}^{j+m-1} \left[\frac{m(1+k)}{m(1+k)+2\bar{\gamma}_b}\right]^i$$
$$\left.\times {}_2F_1\left(\frac{i+1}{2}, \frac{i+2}{2}; 1; \frac{2\bar{\gamma}_b^2}{[m(1+k)+2\bar{\gamma}_b]^2}\right)\right] \tag{6.55}$$

It is quite evident that the resultant error expression is in closed-form, and contains finite series of common functions.

Next, we investigate some special cases of the result presented in (6.55). When $L = 1$ and $m =$ any integer, from (6.55) we get back (6.20), which can also be derived by directly averaging the CEP (4.42) over Nakagami PDF (3.54) applying (A.4). Further, by putting $m = 1$ in (6.55) and using [8, (9.121.1)] we get back (4.47) or (4.48).

In Fig. 6.4, BEP of differentially detected $\pi/4$-DQPSK operating over Nakagami fading channels with SC diversity is depicted for single ($L = 1$) and multichannel ($L = 2$, 4) reception for various integer fading severity parameter values m ($m = 1$, 2, and 4) using (6.55). Specifically, $m = 1$ refers to the Rayleigh fading case given in (4.48) and (4.40).

In Fig. 6.5, the simulated points are superimposed (shown by asterisk mark) on the analytical values obtained in the current section. The simulation and analytical values almost coincide with each other which validate the theoretical analysis.

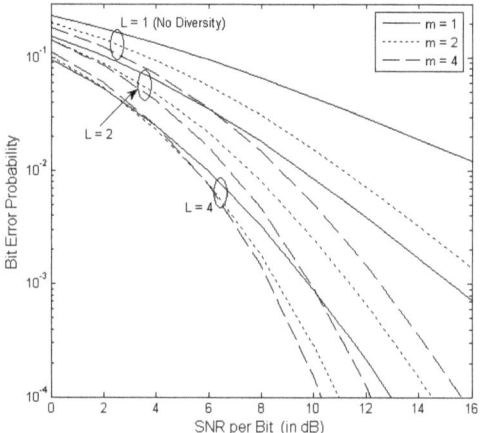

Figure 6.4: BEP of $\pi/4$-DQPSK in Nakagami fading with SC.

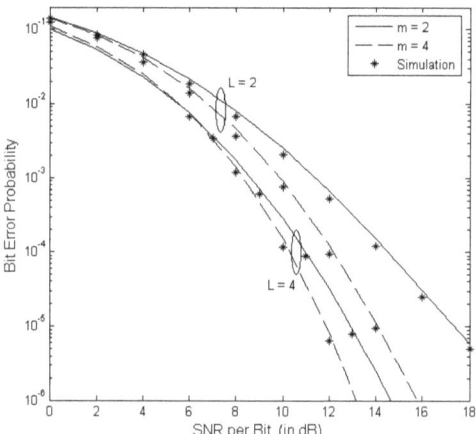

Figure 6.5: Analytical and simulated BEP of $\pi/4$-DQPSK in Nakagami fading with SC.

6.3 Performance of Coherent MFSK

The average SEP of coherent MFSK over Nakagami fading channel may be found by substituting $G_\gamma(s)$ from (3.51) in (4.64) to yield

$$
\begin{aligned}
P_e &= \frac{M-1}{\pi} \int_0^{\pi/2} f_8(\theta) d\theta \\
&\quad - \frac{(M-1)(M-2)}{2\pi} \int_0^{\pi/4} f_9(\theta) f_{10}(\theta) d\theta
\end{aligned}
$$

174

where

$$f_8(\theta) = \left(\frac{2m \sin^2 \theta}{\bar{\gamma} + 2m \sin^2 \theta} \right)^m \qquad (6.57)$$

$$f_9(\theta) = \frac{\sin \theta}{\sqrt{1 + \sin^2 \theta}} \qquad (6.58)$$

$$f_{10}(\theta) = \left[\frac{m(1 + \sin^2 \theta)}{\bar{\gamma} + m(1 + \sin^2 \theta)} \right]^m \qquad (6.59)$$

and $f_2(\theta)$ is defined in (2.112). Like Rician fading, the SEP expression in (6.56) also reduces to (4.65) for a particular value of the fading parameter ($m = 1$). Again, from the generalized error expression $\int_0^\infty P_e(\gamma) f_\gamma(\gamma) d\gamma$ and the Nakagami PDF in (3.54), we find that the averaging requires solution of integrals containing powers and exponential of the dependent variable. Now using the identity, $\int_0^\infty z^{n-1} \exp(-\beta z) dz = \Gamma(n)/\beta^n$, and performing the same integration for all five terms in (4.63), we finally get (6.56) once again.

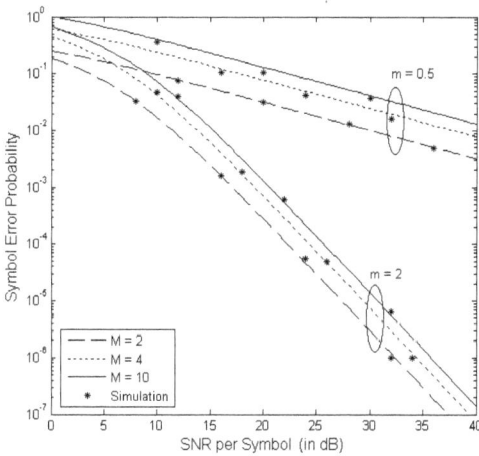

Figure 6.6: SEP of coherent MFSK in Nakagami fading.

Fig. 6.6 plots the analytical results from (6.56) giving symbol error rates for coherent MFSK over Nakagami fading channel. Two different fading parameters ($m = 0.5$ and 2.0) are considered.

6.3.1 SEP for MRC

For the Nakagami channel the SEP is

$$
\begin{aligned}
P_e = {} & \frac{M-1}{\pi} \int_0^{\pi/2} g_5(\theta) d\theta \\
& - \frac{(M-1)(M-2)}{2\pi} \int_0^{\pi/4} g_6(\theta) d\theta \\
& + \frac{(M-1)(M-2)(M-3)}{6\pi^2} \int_0^{\pi/6} g_6(\theta) f_2(\theta) d\theta \\
& + \frac{(M-1)(M-2)(M-3)}{12\pi^2} \int_0^{\sin^{-1}\left(\frac{1}{\sqrt{3}}\right)} g_6(\theta)[\pi - f_2(\theta)] d\theta \\
& - \frac{(M-1)(M-2)(M-3)(M-4)}{24\pi^2} \int_0^{\pi/6} g_6(\theta) f_2(\theta) d\theta
\end{aligned}
\tag{6.60}
$$

which was obtained by substituting the MGF for Nakagami channel in (4.68). In (6.60)

$$
g_5(\theta) = \left(\frac{2m \sin^2 \theta}{\bar{\gamma} + 2m \sin^2 \theta} \right)^{Lm}
\tag{6.61}
$$

$$
g_6(\theta) = \frac{\sin \theta}{\sqrt{1 + \sin^2 \theta}} \left[\frac{m(1 + \sin^2 \theta)}{\bar{\gamma} + m(1 + \sin^2 \theta)} \right]^{Lm}
\tag{6.62}
$$

As a double check, it can be easily verified that putting $L = 1$ in (6.60), we get back (6.56).

Fig. 6.7 shows the SEP for coherent MFSK over Nakagami fading channel with MRC diversity. Keeping the constellation size ($M = 8$) and fading parameter ($m = 2$) fixed, SEP values are depicted for various diversity orders ($L = 2$, 3, and 4).

Chapter Summary

In this chapter, closed-form BEP expressions for $\pi/4$-DQPSK with SC and MRC over Nakagami fading channel are derived. Both the cases, i.e. when the fading parameter m is integer or a fraction, are discussed. For integer fading parameter values, closed-form expressions are attained, while for fractional values the result is presented in an infinite series form. The end results are free from any approximations and do not involve Appell or Lauricella functions, which are difficult to evaluate with common mathematical software packages. Performance of coherent MFSK with MRC is also included in the chapter. Calculations for all other modulation schemes are either available in the open literature or can be easily derived, and therefore not included in this chapter.

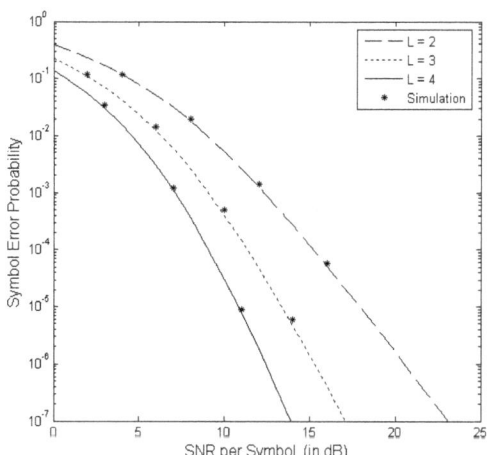

Figure 6.7: SEP of coherent MFSK ($M - 8$) in Nakagami ($m - 2$) fading with MRC.

Chapter 7

Diversity Combining in Hoyt Fading

In satellite links subjected to strong ionospheric scintillation, the amplitude attenuation or fading typically follows Hoyt distribution. Also when mobile satellite channels are modeled as two-state (good/ bad) processes, Hoyt distribution is used to characterize the bad state, while Rician distribution accounts for the good state. Hoyt distribution, also known as Nakagami-q distribution (q being the fading severity parameter), allows us to span the range of fading distribution from one-sided Gaussian ($q = 0$) to Rayleigh fading ($q = 1$), and is used extensively for modeling multipath wireless channels. Despite all these facts, literature pertaining to Hoyt fading model is scarce, and only recently there is a renewed interest in error performance analysis of digital signals over Hoyt fading channel.

The current chapter begins with a brief review of BEP calculation for binary modulations, as given in Section 7.1. Next, in Section 7.2, we derive a closed-form expression for the CDF of the instantaneous SNR in Hoyt channel. The derived CDF is also compared with simulated CDF for identical parameter values. The performance of $\pi/4$-DQPSK in Hoyt channel is discussed in Section 7.3. This is followed by SEP analysis of coherent and non-coherent MFSK, in Section 7.4 and Section 7.5 respectively. Coherent MRC combining is assumed for the coherent MFSK case, while in case of ncMFSK, SC diversity is used at the receiver to avoid phase estimation. Finally, in Section 7.6, we derive MGF, average SNR, outage probability, and SEP for a switched combiner operating over Hoyt fading channel.

7.1 Performance of Binary Modulations

A unified expression for binary modulations in terms of finite integral may be found following the approach in Section 5.1, and is explained by Alouini and Simon [150] in detail. Although, a closed-form BEP expression covering all four basic binary modulations is not available, individual expressions for coherent and non-coherent modulations exist.

From Radaydeh's paper [193], the BEP of coherent binary modulations over Hoyt fading channel may be expressed in terms of the Appell's hypergeometric function of first kind as

$$P_{e,b} = \frac{1}{4\sqrt{1 + 2\alpha\bar{\gamma}_b + p\alpha^2\bar{\gamma}_b^2}}$$
$$\times F_1\left(\frac{1}{2}; \frac{1}{2}, \frac{1}{2}; 2; \frac{1}{1 + m\alpha\bar{\gamma}_b}, \frac{1}{1 + n\alpha\bar{\gamma}_b}\right) \tag{7.1}$$

where $\alpha = 1$ for BPSK and $\alpha = 1/2$ for BFSK. Further, the parameters m and n are defined as, $\{m, n\} = 1 \pm \sqrt{1 - p}$

The calculations for the non-coherent case is rather easy. Integrating the CEP $P_{e,b}(\gamma_b) = 1/2 \exp(-\alpha\gamma_b)$ over the Hoyt PDF given by (3.60) one obtain

$$P_{e,b} = \frac{1}{2\sqrt{1 + 2\alpha\bar{\gamma}_b + p\alpha^2\bar{\gamma}_b^2}} \tag{7.2}$$

where the result in (B.7) is used to perform the integration. In (7.2), the value of α is 1 for DPSK and 1/2 for ncBFSK.

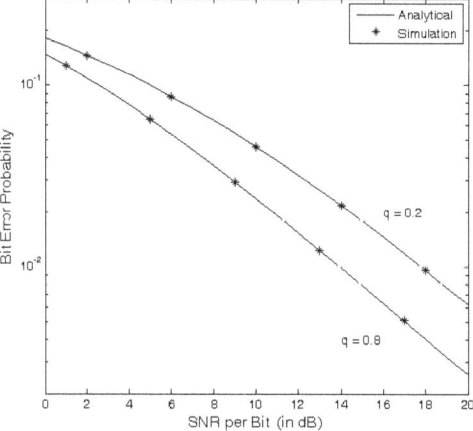

Figure 7.1: BEP of BPSK in Hoyt fading.

Fig. 7.1 shows a plot of (7.1) with $\alpha = 1$, i.e. for BPSK with two different fading parameters. The simulation results are also superimposed on the analytical curves, showing a close fit.

7.2   CDF for Hoyt Fading

When we have indefinite integrals of the type $\int \exp(-ax^2)I_\nu(bx)dx$, they can be expressed with Marcum's Q function from the basic definition

$$Q_m(a,b) = \int_b^\infty x \left(\frac{x}{a}\right)^{m-1} \exp\left(-\frac{a^2+x^2}{2}\right) I_{m-1}(ax)dx \qquad (7.3)$$

which is often very useful for Rician (Nakagami-n) channel calculations. However there exists no direct formula for the integrals $\int \exp(-ax)I_0(bx)dx$, often encountered in Hoyt fading channel calculations.

Let us define, $a = (p^2 + q^2)/2$ and $b = pq$ which give

$$2(a+b) = (q+p)^2 \Rightarrow q+p = \sqrt{2(a+b)} \qquad (7.4a)$$
$$2(a-b) = (q-p)^2 \Rightarrow q-p = \sqrt{2(a-b)} \qquad (7.4b)$$

Adding the two,

$$\begin{aligned}
q &= \sqrt{\frac{a+b}{2}} + \sqrt{\frac{a-b}{2}} \\
&= \sqrt{\frac{a+b}{2} + \frac{a-b}{2} + 2\sqrt{\frac{a^2-b^2}{4}}} \qquad (7.5) \\
&= \sqrt{a + \sqrt{a^2-b^2}}
\end{aligned}$$

while subtracting second from the first,

$$\begin{aligned}
p &= \sqrt{\frac{a+b}{2}} - \sqrt{\frac{a-b}{2}} \\
&= \sqrt{a - \sqrt{a^2-b^2}} \qquad (7.6)
\end{aligned}$$

Also

$$q^2 - p^2 = \sqrt{2(a+b)2(a-b)} = 2\sqrt{a^2-b^2} \qquad (7.7)$$

Expressing the integrand of the generic form in terms of p, q

$$\exp(-ax)I_0(bx)$$
$$= \exp\left(-x\frac{p^2+q^2}{2}\right)I_0(pqx)$$
$$= -\frac{1}{q^2\sqrt{x}}\left[q\sqrt{x}\exp\left(-x\frac{p^2+q^2}{2}\right)\{pI_1(pqx)-qI_0(pqx)\}\right]$$
$$\quad +\frac{1}{q^2\sqrt{x}}q\sqrt{x}\exp\left(-x\frac{p^2+q^2}{2}\right)pI_1(pqx)$$
$$= -\frac{2}{q^2}\left[q\sqrt{x}\exp\left(-x\frac{p^2+q^2}{2}\right)\{pI_1(pqx)-qI_0(pqx)\}\right]\frac{1}{2\sqrt{x}}$$
$$\quad +\frac{1}{q^2}\left[\exp\left(-x\frac{p^2+q^2}{2}\right)pqI_1(pqx)\right.$$
$$\quad \left.-\frac{p^2+q^2}{2}\exp\left(-x\frac{p^2+q^2}{2}\right)I_0(pqx)\right]$$
$$\quad +\frac{p^2+q^2}{2q^2}\exp\left(-x\frac{p^2+q^2}{2}\right)I_0(pqx)$$
$$= -\frac{2}{q^2}\frac{d}{dx}Q_1\left(p\sqrt{x},q\sqrt{x}\right)+\frac{1}{q^2}\frac{d}{dx}\left[\exp\left(-x\frac{p^2+q^2}{2}\right)I_0(pqx)\right]$$
$$\quad +\frac{p^2+q^2}{2q^2}\exp\left(-x\frac{p^2+q^2}{2}\right)I_0(pqx)$$

(7.8)

where we have used $I_0'(z) = I_1(z)$ [9, (9.6.27)] and

$$\frac{d}{dx}Q(ax,bx) = bx\exp\left(-x^2\frac{a^2+b^2}{2}\right)\left[aI_1(abx^2)-bI_0(abx^2)\right] \qquad (7.9)$$

from [194]. In terms of a, b

$$\exp(-ax)I_0(bx)\sqrt{a^2-b^2}$$
$$= \frac{d}{dx}\left[\exp(-ax)I_0(bx)\right.$$
$$\quad \left.-2Q_1\left(\sqrt{a-\sqrt{a^2-b^2}}\sqrt{x},\sqrt{a+\sqrt{a^2-b^2}}\sqrt{x}\right)\right] \qquad (7.10)$$

which means,

$$\int \exp(-ax)I_0(bx)dx$$
$$= \frac{1}{\sqrt{a^2-b^2}}\left[\exp(-ax)I_0(bx)-2Q_1(u,v)\right]+C \qquad (7.11)$$

where $\{u, v\} = \sqrt{(a \mp \sqrt{a^2 - b^2})x}$ and C is the integration constant. This derivation was inspired from [195, (5.7)-(5.8)] by noting the similarity of the Hoyt PDF and the PDF of sum of independent central chi-square random variables of single order.

Using the result derived above, the corresponding CDF may be found by integrating the PDF given in (3.60) over γ as

$$F_\gamma(\gamma) = 1 + \exp\left(-\frac{\gamma}{p\bar{\gamma}}\right) I_0 \left(\frac{\gamma\sqrt{1-p}}{p\bar{\gamma}}\right) - 2Q_1(m_1, n_1) \quad ; \gamma \geq 0 \tag{7.12}$$

where $\{m_1, n_1\} = \sqrt{(1 \mp \sqrt{p})\gamma/(p\bar{\gamma})}$. The constant term ($C = 1$) is found by noting the boundary values $F_\gamma(0) = 0$, $I_0(0) = Q_1(0, 0) = 1$. In terms of the fading parameter q, the CDF is

$$\begin{aligned}
F_\gamma(\gamma) &= 1 + \exp\left[-\frac{(1+q^2)^2\gamma}{4q^2\bar{\gamma}}\right] I_0\left[\frac{(1-q^4)\gamma}{4q^2\bar{\gamma}}\right] \\
&\quad - 2Q_1\left[\sqrt{\frac{(1-q)^2(1+q^2)\gamma}{4q^2\bar{\gamma}}}, \sqrt{\frac{(1+q)^2(1+q^2)\gamma}{4q^2\bar{\gamma}}}\right]
\end{aligned} \tag{7.13}$$

It may be noted that this CDF expression of Hoyt fading SNR in closed-form also was simultaneously derived by Paris [196]. However, the same problem was dealt in this book in an independent manner.

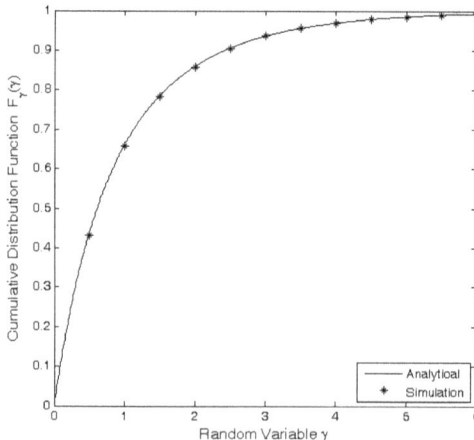

Figure 7.2: Analytical and simulated CDF values for Hoyt ($q = 0.5$) fading SNR.

The expression for CDF of instantaneous SNR over Hoyt fading channel given in (7.12) has been plotted for $q = 0.5$ in Fig. 7.2 alongwith the simulated values. Both the curves give identical values which validates our analysis. As a double check, one may differentiate (7.12) with respect to γ using [9, (9.6.27)] and (7.9) to get back (3.60).

7.3 Performance of $\pi/4$-DQPSK

Following the infinite series representation of modified Bessel functions [9, (9.6.12)]

$$I_0(z) = \sum_{n=0}^{\infty} \frac{z^{2n}}{4^n (n!)^2} \tag{7.14}$$

the Hoyt fading PDF in (3.60) can be expressed as

$$f_\gamma(\gamma_b) = \frac{1}{\sqrt{p}\bar{\gamma}_b} \sum_{n=0}^{\infty} \frac{1}{(n!)^2} \left(\frac{1-p}{4p^2\bar{\gamma}_b^2}\right)^n \gamma_b^{2n} \exp\left(-\frac{\gamma_b}{p\bar{\gamma}_b}\right) \quad ; \gamma_b \geq 0 \tag{7.15}$$

Now the BEP of $\pi/4$-DQPSK over Hoyt fading channel may be found when the CEP given in (4.42) is averaged over (7.15). Noting that $\int_0^\infty f_\gamma(\gamma_b)d\gamma_b = 1$ and from the integration result provided in (A.4), the final average BEP for Hoyt fading becomes

$$
\begin{aligned}
P_{e,b} =\ & \frac{1}{2}\Bigg[1 - \frac{\sqrt{2}p^{3/2}\bar{\gamma}_b}{1+2p\bar{\gamma}_b} \sum_{n=0}^{\infty} \binom{2n}{n}\left(\frac{1-p}{4}\right)^n \\
& \times \sum_{i=0}^{2n} \frac{1}{(1+2p\bar{\gamma}_b)^i} \, {}_2F_1\left(\frac{i+1}{2}, \frac{i+2}{2}; 1; \frac{2p^2\bar{\gamma}_b^2}{(1+2p\bar{\gamma}_b)^2}\right)\Bigg]
\end{aligned}
\tag{7.16}
$$

which is, of course, a series solution. Graphical analysis of $\pi/4$-DQPSK scheme over Hoyt fading channel is shown in Fig. 7.3 using (7.15) with three different fading parameter values ($q = 0.20,\ 0.35,$ and 0.80).

An alternative finite integral solution for the same may be obtained when the CEP given in the form of finite integral (see (4.50)) is averaged over the basic PDF mentioned in (3.60). The SEP evaluation in this method requires the following double integral to be solved

$$
\begin{aligned}
P_{e,b} =\ & \frac{1}{2\pi\sqrt{p}\bar{\gamma}_b} \int_0^\pi \frac{1}{\sqrt{2}-\cos\theta} \int_0^\infty I_0\left(\frac{\gamma_b\sqrt{1-p}}{p\bar{\gamma}_b}\right) \\
& \times \exp\left[-\gamma_b\left(\frac{1+p\bar{\gamma}_b(2-\sqrt{2}\cos\theta)}{p\bar{\gamma}_b}\right)\right] d\gamma_b d\theta
\end{aligned}
\tag{7.17}
$$

where we have already performed a change of order of integration. To solve the inner integral, we take help of the identity (B.8). With the help of the integral identity, the

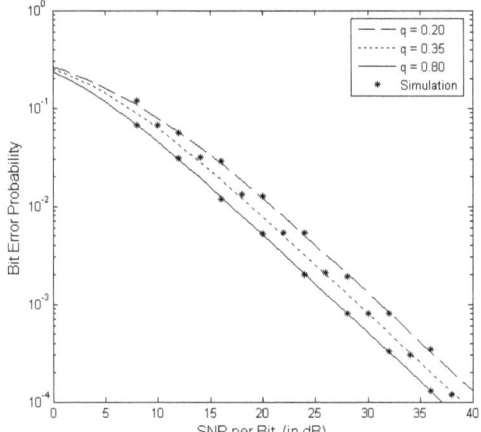

Figure 7.3: BEP of $\pi/4$-DQPSK in Hoyt fading.

final error expression can be simplified as

$$P_{e,b} = \frac{1}{2\pi} \int_0^\pi \frac{\left[1 + \bar{\gamma}_b(2 - \sqrt{2}\cos\theta)\{2 + p\bar{\gamma}_b(2 - \sqrt{2}\cos\theta)\}\right]^{-\frac{1}{2}}}{\sqrt{2} - \cos\theta} d\theta \qquad (7.18)$$

7.4 Performance of Coherent MFSK

For Hoyt fading the SEP takes the form

$$
\begin{aligned}
P_e &= \frac{M-1}{\pi} \int_0^{\pi/2} f_{11}(\theta)d\theta \\
&\quad - \frac{(M-1)(M-2)}{2\pi} \int_0^{\pi/4} f_9(\theta)f_{12}(\theta)d\theta \\
&\quad + \frac{(M-1)(M-2)(M-3)}{6\pi^2} \int_0^{\pi/6} f_9(\theta)f_2(\theta)f_{12}(\theta)d\theta \\
&\quad + \frac{(M-1)(M-2)(M-3)}{12\pi^2} \int_0^{\sin^{-1}\left(\frac{1}{\sqrt{3}}\right)} f_9(\theta)[\pi - f_2(\theta)]f_{12}(\theta)d\theta \\
&\quad - \frac{(M-1)(M-2)(M-3)(M-4)}{24\pi^2} \int_0^{\pi/6} f_9(\theta)f_2(\theta)f_{12}(\theta)d\theta
\end{aligned}
\qquad (7.19)
$$

which was obtained by substituting $G_\gamma(s)$ from (3.62) in (4.64). In (7.19)

$$f_{11}(\theta) = \left[1 + \frac{\bar{\gamma}}{\sin^2\theta} + \left\{\frac{q\bar{\gamma}}{(1+q^2)\sin^2\theta}\right\}\right]^{-1/2} \tag{7.20}$$

$$f_{12}(\theta) = \left[1 + \frac{2\bar{\gamma}}{1+\sin^2\theta} + \left\{\frac{2q\bar{\gamma}}{(1+q^2)(1+\sin^2\theta)}\right\}\right]^{-1/2} \tag{7.21}$$

and $f_2(\theta)$, $f_9(\theta)$ are defined in (2.112) and (6.58) respectively. As a check, one may verify that for $q = 1$, (7.19) reduces to (4.65). SEP calculation in Hoyt fading through PDF method is similar to the Rician case in that, it also involves integrations containing exponential and Bessel functions but of the form $\int_0^\infty \exp(-ax)I_0(bx)dx$. Solution to such kind of integral may be found in (B.7).

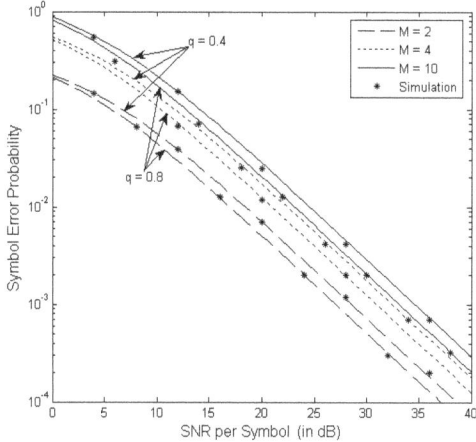

Figure 7.4: SEP of coherent MFSK in Hoyt fading.

Analysis of coherent MFSK schemes over Hoyt (Nakagami-q) fading channel is shown in Fig. 7.4 using (7.19) for two different fading parameters (0.4 and 0.8), and for various values of modulation order M (=2, 4, and 10).

For coherent MFSK, instead of (2.110) if we assume the crude upper bound (2.113) [2] and calculate the SEP for fading channels based on it, we would end up with the first terms present in (4.65), (5.33), (6.56) or (7.19). Fig. 7.5 compares these bounds with those presented in this book. A quick examination reveals that these bounds are not only far away from the actual values (simulated points), but also provide us with unrealistic values under some circumstances. For example, in the low SNR region the SEP predicted by these crude bounds may be more than 1, i.e. more number of erroneous symbols than

185

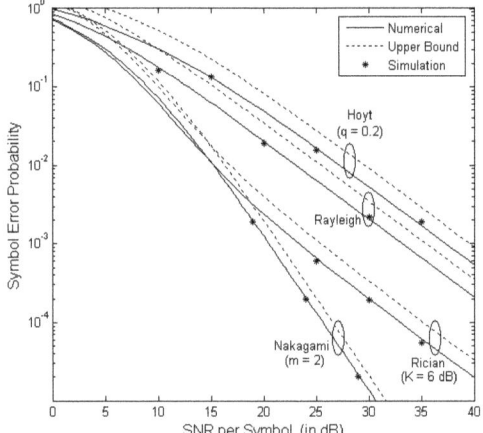

Figure 7.5: Comparison of the proposed approximation (numerical) and conventional approximation (upper bound) with the simulated SEP of coherent MFSK ($M = 8$) for different fading channels.

actually transmitted.

7.4.1 SEP for MRC

Like Rayleigh, Rician, or Nakagami fading, the SEP for Hoyt fading with MRC diversity channel also can be found easily by substituting the corresponding MGF in (4.68)

$$
\begin{aligned}
P_e &= \frac{M-1}{\pi} \int_0^{\pi/2} g_7(\theta) d\theta \\
&\quad - \frac{(M-1)(M-2)}{2\pi} \int_0^{\pi/4} g_8(\theta) d\theta \\
&\quad + \frac{(M-1)(M-2)(M-3)}{6\pi^2} \int_0^{\pi/6} g_8(\theta) f_2(\theta) d\theta \\
&\quad + \frac{(M-1)(M-2)(M-3)}{12\pi^2} \int_0^{\sin^{-1}\left(\frac{1}{\sqrt{3}}\right)} g_8(\theta)[\pi - f_2(\theta)] d\theta \\
&\quad - \frac{(M-1)(M-2)(M-3)(M-4)}{24\pi^2} \int_0^{\pi/6} g_8(\theta) f_2(\theta) d\theta
\end{aligned}
\tag{7.22}
$$

where

$$g_7(\theta) = \left[1 + \frac{\bar{\gamma}}{\sin^2\theta} + \left\{\frac{q\bar{\gamma}}{(1+q^2)\sin^2\theta}\right\}\right]^{-L/2} \qquad (7.23)$$

$$g_8(\theta) = f_9(\theta)\left[1 + \frac{2\bar{\gamma}}{1+\sin^2\theta} + \left\{\frac{2q\bar{\gamma}}{(1+q^2)(1+\sin^2\theta)}\right\}\right]^{-L/2} \qquad (7.24)$$

where $f_9(\theta)$ is defined in (6.58).

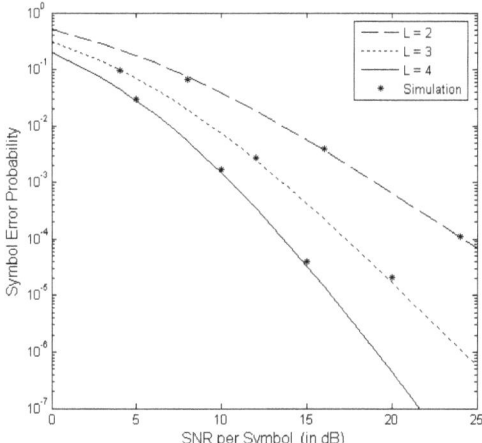

Figure 7.6: SEP of coherent MFSK ($M = 8$) in Hoyt fading with MRC.

In Fig. 7.6, keeping the constellation size ($M = 8$) and fading parameter ($q = 0.4$) fixed, SEP values are depicted for various diversity orders ($L - 2$, 3, and 4).

Fig. 7.7 compares the upper bound derived from conventional approximation with the one presented in this book, and it shows that the nature of the curves is entirely different for two approximations in low SNR values.

7.5 Performance of Non-coherent MFSK

In [193], Radaydeh derived SEP of non-coherent correlated binary signals in Hoyt fading channels in terms of the Lauricella's hypergeometric function, $F_D^{(n)}; n \geq 1$, which can be evaluated numerically using its integral or converging series representation. However, the error rate for ncMFSK can be calculated in simple closed-form by substituting the Hoyt PDF and the CEP for non-coherent MFSK given in (3.60) and (5.36) respectively in the

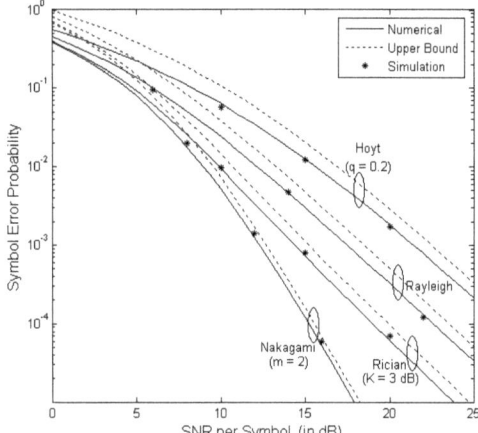

Figure 7.7: Comparison of the proposed approximation (numerical) and conventional approximation (upper bound) with the simulated SEP of coherent MFSK ($M = 8$) for different fading channels with MRC ($L = 2$) diversity.

generic equation $\int_0^\infty P_e(\gamma) f_\gamma(\gamma) d\gamma$ to obtain

$$
\begin{aligned}
P_e ={}& \frac{1}{\sqrt{p\bar{\gamma}}} \sum_{j=1}^{M-1} \frac{(-1)^{j+1}}{j+1} \binom{M-1}{j} \\
&\times \int_0^\infty \exp\left[-\gamma\left(\frac{j}{j+1} + \frac{1}{p\bar{\gamma}}\right)\right] I_0\left(\frac{\gamma\sqrt{1-p}}{p\bar{\gamma}}\right) d\gamma
\end{aligned}
\tag{7.25}
$$

where the integral in (7.25) may be solved using (B.7). After some algebraic manipulation we get

$$
P_e = \sum_{j=1}^{M-1} \binom{M-1}{j} \frac{(-1)^{j+1}}{\sqrt{(1+j+j\bar{\gamma})^2 - (1-p)(j\bar{\gamma})^2}}
\tag{7.26}
$$

which reduces to (5.39) for the Rayleigh fading case ($p = 1$).

Fig. 7.8 shows a plot of (7.26) for two different fading parameters ($q = 0.2$ and 0.8) and for two different modulation orders ($M = 2$ and 4). The trend of SEP to increase with modulation order and fading severity (denoted by a decrement of q) is clearly visible.

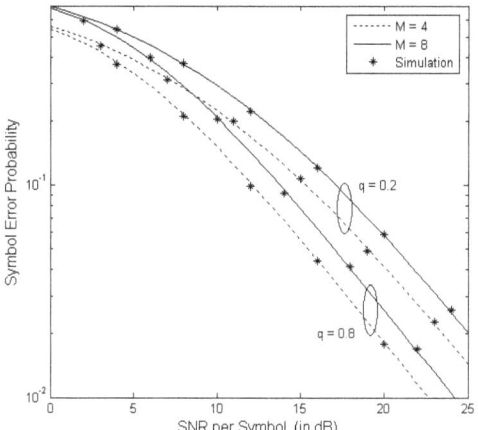

Figure 7.8: SEP of ncMFSK in Hoyt fading.

7.5.1 SEP for SC

The performance of EGC [197, 198] and MRC [198] over Hoyt fading channel had been investigated by many researchers. In particular, Radaydeh and Matalgah [197] derived the BEP of ncMFSK signals over multichannel non-identically distributed Hoyt fading employing post-detection non-coherent EGC, whereas Fraidenraich et al. [198] described the second-order statistics with MRC and EGC. Recently, in a paper by Krstić et al. [199], the error performance with generalized selection combining (GSC) over Hoyt channels was examined, and the authors presented numerical results for BPSK. In this subsection we derive the SEP of ncMFSK with SC in Hoyt fading channel by mimicking Haghani and Beaulieu's approach [152]. We start with the modified SEP formula

$$P_e = -\int_0^{\pi/2} P_e'(\tan\theta) \left[F_\gamma(\tan\theta)\right]^L \sec^2\theta d\theta \tag{7.27}$$

as given in (5.41) and insert the expressions of $P_e'(\tan\theta)$ from (5.42) as well as for $F_\gamma(\tan\theta)$ as derived in (7.12), to obtain

189

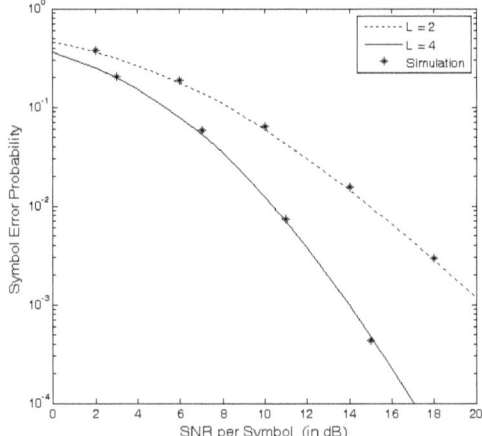

Figure 7.9: SEP of ncMFSK ($M = 4$) in Hoyt ($q = 0.5$) fading with SC.

$$P_e = \sum_{j=1}^{M-1} \frac{j(-1)^{j+1}}{(j+1)^2} \binom{M-1}{j} \int_0^{\pi/2} \exp\left(-\frac{j \tan \theta}{j+1}\right)$$
$$\times \left[1 + \exp\left(-\frac{\tan \theta}{p\bar{\gamma}}\right) I_0\left(\frac{\tan \theta \sqrt{1-p}}{p\bar{\gamma}}\right)\right.$$
$$\left. -2Q\left(\sqrt{\frac{(1-\sqrt{p})\tan \theta}{p\bar{\gamma}}}, \sqrt{\frac{(1+\sqrt{p})\tan \theta}{p\bar{\gamma}}}\right)\right]^L \sec^2 \theta d\theta \tag{7.28}$$

In Fig. 7.9, SEP for nc4FSK in Hoyt fading channel with SC are shown for two different diversity orders ($L = 2$ and 4), keeping the fading parameter q fixed at 0.5. As expected, the SEP performance significantly improves when more number of diversity branches are introduced at receiver.

7.6 Performance of Switched Diversity

7.6.1 MGF of SSC output SNR

Assuming IID branches $\bar{\gamma}_1 = \bar{\gamma}_2 = \bar{\gamma}$, the PDF of the output SNR for SSC is given by (3.99). Averaging $\exp(-s\gamma)$ over this PDF, the corresponding MGF of γ may be calculated as

$$
\begin{aligned}
M_{\gamma_{SSC}}(s) &= \int_0^\infty f_{\gamma_{SSC}}(\gamma)\exp(-s\gamma)d\gamma \\
&= F_\gamma(\gamma_T)\int_0^{\gamma_T} f_\gamma(\gamma)\exp(-s\gamma)d\gamma \\
&\quad +[1+F_\gamma(\gamma_T)]\int_{\gamma_T}^\infty f_\gamma(\gamma)\exp(-s\gamma)d\gamma \\
&= [1+F_\gamma(\gamma_T)]\int_0^\infty f_\gamma(\gamma)\exp(-s\gamma)d\gamma \\
&\quad -\int_0^{\gamma_T} f_\gamma(\gamma)\exp(-s\gamma)d\gamma \\
&= [1+F_\gamma(\gamma_T)]M_\gamma(s)-\mathcal{I}
\end{aligned}
\tag{7.29}
$$

where $f_\gamma(\gamma)$, $F_\gamma(\gamma)$, and $M_\gamma(s)$ are the PDF, CDF, and MGF of single Hoyt fading channel defined in (3.60), (7.12), and (3.62) respectively and

$$
\begin{aligned}
\mathcal{I} &= \int_0^{\gamma_T} f_\gamma(\gamma)\exp(-s\gamma)d\gamma \\
&= \frac{1}{\sqrt{p\bar\gamma}}\int_0^{\gamma_T}\exp\left[-\gamma\left(s+\frac{1}{p\bar\gamma}\right)\right]I_0\left(\frac{\gamma\sqrt{1-p}}{p\bar\gamma}\right)d\gamma
\end{aligned}
\tag{7.30}
$$

Using the result in (7.11), noting that $\exp(0)=I_0(0)=Q_1(0,0)=1$, and from the definition of $M_\gamma(s)$ the integral $\mathcal{I}$ may be evaluated as

$$
\begin{aligned}
\mathcal{I} &= M_\gamma(s)\left\{1+\exp\left[-\gamma_T\left(s+\frac{1}{p\bar\gamma}\right)\right]\right. \\
&\qquad \left. \times I_0\left(\frac{\gamma_T\sqrt{1-p}}{p\bar\gamma}\right)-2Q_1(m_2,n_2)\right\}
\end{aligned}
\tag{7.31}
$$

where

$$
\{m_2,n_2\}=\sqrt{\frac{\gamma_T}{p\bar\gamma}\left[1+sp\bar\gamma\mp\frac{\sqrt{p}}{M_\gamma(s)}\right]}
\tag{7.32}
$$

Putting the value of $\mathcal{I}$ from (7.31) in (7.29), ultimately the MGF

$$
\begin{aligned}
M_{\gamma_{SSC}}(s) &= M_\gamma(s)\left\{F_\gamma(\gamma_T)-\exp\left[-\gamma_T\left(s+\frac{1}{p\bar\gamma}\right)\right]\right. \\
&\qquad \left. \times I_0\left(\frac{\gamma_T\sqrt{1-p}}{p\bar\gamma}\right)+2Q_1(m_2,n_2)\right\}
\end{aligned}
\tag{7.33}
$$

7.6.2 Average SNR with SSC

The average SNR at the SSC output can be obtained by averaging γ over $f_{\gamma_{SSC}}(\gamma)$ yielding

$$
\begin{aligned}
\bar{\gamma}_{SSC} &= \int_0^\infty \gamma f_{\gamma_{SSC}}(\gamma)d\gamma \\
&= [1 + F_\gamma(\gamma_T)]\int_0^\infty \gamma f_\gamma(\gamma)d\gamma - \int_0^{\gamma_T} \gamma f_\gamma(\gamma)d\gamma
\end{aligned}
\tag{7.34}
$$

The first integral in (7.34) is nothing but $\bar{\gamma}$, while the second one may be evaluated using integration by parts. Rearranging the terms we get

$$
\begin{aligned}
\bar{\gamma}_{SSC} &= [1 + F_\gamma(\gamma_T)]\bar{\gamma} - \left[\{\gamma F_\gamma(\gamma)\}|_{\gamma=0}^{\gamma_T} - \int_0^{\gamma_T} F_\gamma(\gamma)d\gamma\right] \\
&= \bar{\gamma} + (\bar{\gamma} - \gamma_T)F_\gamma(\gamma_T) + \int_0^{\gamma_T} F_\gamma(\gamma)d\gamma
\end{aligned}
\tag{7.35}
$$

From (7.12)

$$
\int_0^{\gamma_T} F_\gamma(\gamma)d\gamma = \gamma_T + \mathcal{I}_1 - 2\mathcal{I}_2
\tag{7.36}
$$

where

$$
\mathcal{I}_1 = \int_0^{\gamma_T} \exp\left(-\frac{\gamma}{p\bar{\gamma}}\right) I_0\left(\frac{\gamma\sqrt{1-p}}{p\bar{\gamma}}\right)d\gamma
\tag{7.37}
$$

and

$$
\mathcal{I}_2 = \int_0^{\gamma_T} Q_1\left(\sqrt{\frac{(1-\sqrt{p})\gamma}{p\bar{\gamma}}}, \sqrt{\frac{(1+\sqrt{p})\gamma}{p\bar{\gamma}}}\right)d\gamma
\tag{7.38}
$$

Inserting (7.36) and the expression for $F_\gamma(\gamma_T)$ in (7.35) we have

$$
\begin{aligned}
\bar{\gamma}_{SSC} &= (\bar{\gamma} + \gamma_T) + (\bar{\gamma} - \gamma_T)\left[1 + \exp\left(-\frac{\gamma_T}{p\bar{\gamma}}\right) I_0\left(\frac{\gamma_T\sqrt{1-p}}{p\bar{\gamma}}\right)\right. \\
&\qquad \left. -2Q_1\left(\sqrt{\frac{(1-\sqrt{p})\gamma_T}{p\bar{\gamma}}}, \sqrt{\frac{(1+\sqrt{p})\gamma_T}{p\bar{\gamma}}}\right)\right] \\
&\quad +\mathcal{I}_1 - 2\mathcal{I}_2
\end{aligned}
\tag{7.39}
$$

The first integral may be readily solved using (7.11)

$$\mathcal{I}_1 = \sqrt{p}\bar{\gamma}\left[1 + \exp\left(-\frac{\gamma_T}{p\bar{\gamma}}\right) I_0\left(\frac{\gamma_T\sqrt{1-p}}{p\bar{\gamma}}\right)\right.$$
$$\left. -2Q_1\left(\sqrt{\frac{(1-\sqrt{p})\gamma_T}{p\bar{\gamma}}}, \sqrt{\frac{(1+\sqrt{p})\gamma_T}{p\bar{\gamma}}}\right)\right] \tag{7.40}$$

and thus

$$\bar{\gamma}_{SSC} = \bar{\gamma}(2 + \sqrt{p})$$
$$+(\bar{\gamma} - \gamma_T + \sqrt{p}\bar{\gamma})\exp\left(-\frac{\gamma_T}{p\bar{\gamma}}\right) I_0\left(\frac{\gamma_T\sqrt{1-p}}{p\bar{\gamma}}\right)$$
$$-2(\bar{\gamma} - \gamma_T + \sqrt{p}\bar{\gamma})Q_1\left(\sqrt{\frac{(1-\sqrt{p})\gamma_T}{p\bar{\gamma}}}, \sqrt{\frac{(1+\sqrt{p})\gamma_T}{p\bar{\gamma}}}\right) \tag{7.41}$$
$$-2\mathcal{I}_2$$

For the second integral, we make use of the following result [195, (B.43)] namely

$$\int_0^c Q_1\left(a\sqrt{z}, b\sqrt{z}\right) dz$$
$$= 2\int_0^{\sqrt{c}} x Q_1(ax, bx) dx$$
$$= cQ_1\left(a\sqrt{c}, b\sqrt{c}\right) + \frac{bc}{a^2 - b^2}\exp\left(-\frac{a^2 + b^2}{2}c\right)$$
$$\times [bI_0(abc) + aI_1(abc)]$$
$$-\frac{2b^2}{|a^2 - b^2|(a^2 - b^2)}\left[1 + \exp\left(-\frac{a^2 + b^2}{2}c\right) I_0(abc)\right.$$
$$\left. -2Q_1\left(\min(a, b)\sqrt{c}, \max(a, b)\sqrt{c}\right)\right] \tag{7.42}$$

to obtain

$$\mathcal{I}_2 = \frac{\bar{\gamma}\left(1 + \sqrt{p}\right)}{2}$$
$$+ [\gamma_T - \bar{\gamma}\left(1 + \sqrt{p}\right)] Q_1\left(\sqrt{\frac{(1 - \sqrt{p})\gamma_T}{p\bar{\gamma}}}, \sqrt{\frac{(1 + \sqrt{p})\gamma_T}{p\bar{\gamma}}}\right)$$
$$+ \frac{1 + \sqrt{p}}{2}\left(\bar{\gamma} - \frac{\gamma_T}{\sqrt{p}}\right)\exp\left(-\frac{\gamma_T}{p\bar{\gamma}}\right) I_0\left(\frac{\gamma_T\sqrt{1-p}}{p\bar{\gamma}}\right) \tag{7.43}$$
$$- \frac{\gamma_T}{2}\left(\sqrt{\frac{1}{p}} - 1\right)\exp\left(-\frac{\gamma_T}{p\bar{\gamma}}\right) I_1\left(\frac{\gamma_T\sqrt{1-p}}{p\bar{\gamma}}\right)$$

after considerable algebraic manipulation. Putting the value of $\mathcal{I}_2$ from (7.43) in (7.41) and simplifying a bit by grouping the similar terms we finally get

$$
\begin{aligned}
\bar{\gamma}_{SSC} = \ & \bar{\gamma} + \frac{\gamma_T}{\sqrt{p}} \exp\left(-\frac{\gamma_T}{p\bar{\gamma}}\right) \\
& \times \left[I_0\left(\frac{\gamma_T\sqrt{1-p}}{p\bar{\gamma}}\right) + \sqrt{1-p}\, I_1\left(\frac{\gamma_T\sqrt{1-p}}{p\bar{\gamma}}\right) \right]
\end{aligned}
\tag{7.44}
$$

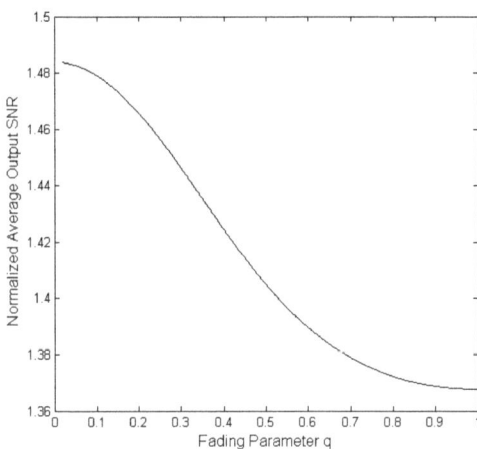

Figure 7.10: Normalized average output SNR as a function of fading parameter (q) for SSC with optimum threshold.

It can be easily shown [200] that the optimum switching threshold for maximizing average combiner output SNR is simply average SNR in each branch, i.e., in order to maximize $\bar{\gamma}_{SSC}$, we should set $\gamma_T^* = \bar{\gamma}$. Under these circumstances the optimum average SNR

$$
\bar{\gamma}_{SSC}^* = \bar{\gamma} + \frac{\bar{\gamma}}{\sqrt{p}} \exp\left(-\frac{1}{p}\right) \left[I_0\left(\frac{\sqrt{1-p}}{p}\right) + \sqrt{1-p}\, I_1\left(\frac{\sqrt{1-p}}{p}\right) \right]
\tag{7.45}
$$

As a check, one may verify that (7.44) and (7.45) reduces to (3.103) and (3.104) respectively for $p = q = 1$, i.e., for the Rayleigh fading case.

Fig. 7.10 plots the normalized average output SNR for SSC diversity ($\bar{\gamma}_{SSC}^*/\bar{\gamma}$) versus the Hoyt fading parameter (q) with optimum threshold. It is quite evident from the figure that the average SNR decreases as the fading severity increases. In other words, SSC diversity provides better SNR gain when the channel conditions are severely degraded by fading.

7.6.3 MGF of SEC output SNR

Assuming IID L branch diversity, the PDF of the output SNR for SEC is given by (3.108). After some algebraic manipulation, the corresponding MGF of γ may be calculated as

$$
\begin{aligned}
M_{\gamma_{SEC}}(s) &= \int_0^\infty f_{\gamma_{SEC}}(\gamma)\exp(-s\gamma)d\gamma \\
&= M_\gamma(s)\sum_{k=0}^{L-1}[F_\gamma(\gamma_T)]^k - \mathcal{I}\sum_{k=0}^{L-2}[F_\gamma(\gamma_T)]^k
\end{aligned}
\tag{7.46}
$$

where $\mathcal{I}$ is defined in (7.31), $F_\gamma(\gamma_T) = F_\gamma(\gamma)|_{\gamma=\gamma_T}$, and $F_\gamma(\gamma)$ and $M_\gamma(s)$ are the CDF and MGF of single Hoyt channel given by (7.12) and (3.62) respectively. Inserting the expression for $\mathcal{I}$ in (7.46), the MGF becomes

$$
\begin{aligned}
M_{\gamma_{SEC}}(s) = M_\gamma(s)\Bigg[& F_\gamma^{L-1}(\gamma_T) - \sum_{k=0}^{L-2}F_\gamma^k(\gamma_T) \\
& \left\{ \exp\left[-\gamma_T\left(s+\frac{1}{p\bar\gamma}\right)\right]\right. \\
& \times I_0\left(\frac{\gamma_T\sqrt{1-p}}{p\bar\gamma}\right) - 2Q_1(m_2,n_2)\bigg\}\Bigg]
\end{aligned}
\tag{7.47}
$$

where $\{m_2, n_2\}$ is defined in (7.32).

7.6.4 Outage Probability with SSC and SEC

The outage probability $P_{o,LC}$ of a linear diversity combiner is defined as the probability that its output SNR γ falls below a given threshold γ_o, $\Pr[\gamma < \gamma_o]$, and therefore, can be obtained from $F_{\gamma_{LC}}(\gamma_o)$, i.e., the CDF expression evaluated at $\gamma = \gamma_o$. Quite naturally for SEC (or SSC), the outage probability $P_{o,SEC}$ is a function of both threshold SNR γ_o and switching threshold γ_T. It is interesting to note that, for $\gamma_T = \gamma_o$ we get the minimum value $(P_{o,SEC}^*)$ for outage probability and the combiner in this condition behaves exactly like SC [73, 200]

$$
P_{o,SEC}^* = \min\left(P_{o,SEC}\right) = \left.P_{o,SEC}\right|_{\gamma_T=\gamma_o} = P_{o,SC}
\tag{7.48}
$$

where

$$
P_{o,SC} = F_{\gamma_{SC}}(\gamma_T) = [F_\gamma(\gamma_T)]^L
\tag{7.49}
$$

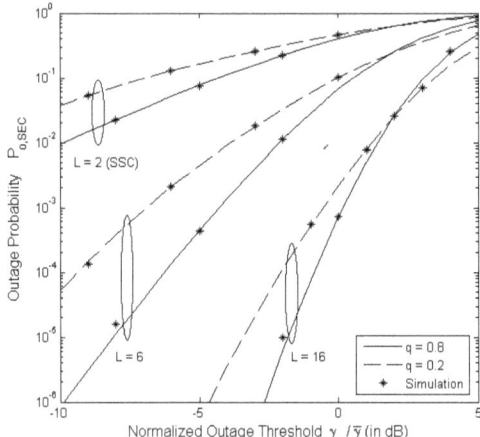

Figure 7.11: Minimum outage probability of SEC versus normalized threshold SNR per symbol ($\hat{\gamma} = \gamma_o/\bar{\gamma}$) in Hoyt fading.

Defining normalized threshold as $\hat{\gamma} = \gamma_o/\bar{\gamma}$ and from (7.12)

$$P_{o,SEC}^* = \left[1 + \exp\left(-\frac{\hat{\gamma}}{p}\right) I_0\left(\frac{\hat{\gamma}\sqrt{1-p}}{p}\right) - 2Q_1(m_3, n_3) \right]^L \tag{7.50}$$

where $\{m_3, n_3\} = \sqrt{(1 \mp \sqrt{p})\hat{\gamma}/p}$. Accordingly, for SSC diversity ($L = 2$) the outage probability is given by

$$P_{o,SSC}^* = \left[1 + \exp\left(-\frac{\hat{\gamma}}{p}\right) I_0\left(\frac{\hat{\gamma}\sqrt{1-p}}{p}\right) - 2Q_1(m_3, n_3) \right]^2 \tag{7.51}$$

Fig. 7.11 depicts theoretical outage probability of an L branch SEC receiver for a range of normalized threshold SNR.

7.6.5 SEP and Optimum Threshold with SEC

The SEP for different family of modulations such as PSK, FSK, and QAM can be attained easily following the MGF based approach discussed in Section 3.2.2. Using the MGF given in (7.47) and the conditional SEP expressions from Chapter 1.3, average SEP for SEC over Hoyt fading channel may be given as

MPSK:

$$P_e = \frac{1}{\pi} \int_0^{\pi-\pi/M} M_{\gamma_{SEC}}[-f_1(\theta)]d\theta \tag{7.52}$$

MDPSK:

$$P_e = \frac{1}{\pi} \int_0^{\pi-\pi/M} M_{\gamma_{SEC}}[-f_2(\theta)]d\theta \tag{7.53}$$

Square MQAM:

$$P_e = \frac{4b}{\pi} \left\{ \int_0^{\pi/2} M_{\gamma_{SEC}}[-f_3(\theta)]d\theta - b \int_0^{\pi/4} M_{\gamma_{SEC}}[-f_3(\theta)]d\theta \right\} \tag{7.54}$$

Non-coherent MFSK:

$$P_e = \frac{1}{M} \sum_{i=2}^{M} (-1)^i \binom{M}{i} M_{\gamma_{SEC}} \left(\frac{1}{i} - 1 \right) \tag{7.55}$$

Coherent MFSK:

$$P_e \leq \frac{M-1}{\pi} \int_0^{\pi/2} M_{\gamma_{SEC}}[-f_4(\theta)]d\theta \tag{7.56}$$

where M is the modulation order, $f_1(\theta) = \sin^2(\pi/M)/\sin^2(\theta)$, $f_2(\theta) = \sin^2(\pi/M)/[1 + \cos(\pi/M)\cos(\theta)]$, $f_3(\theta) = a/\sin^2(\theta)$, $a = (1.5)/(M-1)$, $b = 1 - 1/\sqrt{M}$, and $f_4(\theta) = 1/[2\sin^2(\theta)]$. The end expressions involve elementary functions and reduce to the well known SSC (dual or mutli-branch) case for $L = 2$ or to the Rayleigh fading case for $q = 1$ (or $p = 1$).

In general, the optimum switching threshold γ_T^* ensuring minimum SEP is computed through the identity $\partial P_e/\partial \gamma_T|_{\gamma_T = \gamma_T^*} = 0$. As an indicative example, we consider dual-branch SEC (or SSC) with non-coherent BFSK modulation. Substituting (7.47) in (7.55), differentiating the same, and putting $L = M = 2$ we get

$$\gamma_T^* = \ln \left(1 + \bar{\gamma} + \frac{p\bar{\gamma}^2}{4} \right) \tag{7.57}$$

However, closed-form expressions for γ_T^* are not always available, and we may have to resort to numerical techniques. Fig. 7.12 provides average SEP of nc8FSK for SEC with optimum switching threshold in Hoyt channel. The optimum threshold values are calculated through numerical minimization techniques in MATLAB.

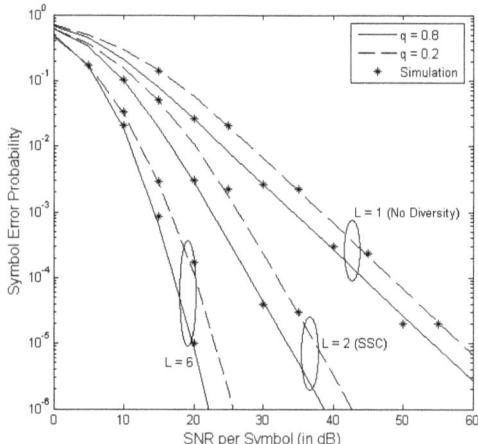

Figure 7.12: SEP of ncMFSK ($M = 8$) for SEC with optimum switching threshold in Hoyt fading.

Chapter Summary

The major contribution of this chapter is the derivation of a closed-form expression for CDF of the instantaneous SNR in Hoyt channel. During the derivation, we have also found the solution of the indefinite integrals of the form $\int \exp(-ax)I_0(bx)dx$. This CDF and associated formulas are then used to find out the error probability of different modulation schemes - $\pi/4$-DQPSK, MFSK, and ncMFSK. BEP/ SEP values for these modulations with diversity are provided in simple integral forms. The switched diversity case is covered in a bit more detail. We have derived the MGF and outage probability expressions with SSC and SEC for Hoyt fading channel. The average SNR with SSC is investigated and an analytical expression for the optimum average SNR is presented. Finally, we computed SEP for all the modulations considered in the book in presence of SEC diversity and provided guidelines to compute the corresponding optimum switching threshold.

Chapter 8

Performance Simulation with MATLAB

Hardware prototyping of wireless systems is becoming more and more expensive and time consuming as the communication systems increase in sophistication and complexity. In fact, it would now usually be impossible to prototype all credible solutions to a given communication problem [3]. With the various analytical models presented in the earlier chapters, a higher layer of abstraction can be achieved. These models, through the development of equations, can predict performance of digital wireless systems under various propagation conditions and design choices. However, the accuracy of such predictions might be questionable in absence of any validation technique. Computer simulation is generally the preferred method for such validations.

We have used MATLAB, a general purpose mathematical tool, to develop the simulation codes. The digital communication system is broken into functional blocks and BEP/ SEP values are simulated through the Monte Carlo procedure. Section 8.1 gives an account of the corresponding block diagram along with some general simulation guidelines. The following three sections, Section 8.2, Section 8.3, and Section 8.4, describe detailed modelling of the modulation/ demodulation process, fading channel, and diversity combiners, respectively.

8.1   Simulation of SEP/ BEP

The most convenient way to simulate SEP or BEP for digital signals is to use Monte Carlo Technique. Monte Carlo is a stochastic simulation process that use random sampling to estimate the output of an experiment [201]. It is a very general method in the sense that unlike the quasi-analytic or other simulation methods, there is neither a linearity restriction on the system, nor any apriori knowledge of the noise/ fading characteristics at the decision block is needed [3]. For estimating SEP (or BEP) of a digital communication system with Monte Carlo procedure, a large number of digital symbols (or bits) are passed through the system and errors are counted at the receiver output. If $Count_{sym}$ number of symbols are processed by the system out of which $Count_{err}$ errors occur, then the estimate of error probability is

$$\hat{P}_e = \frac{Count_{err}}{Count_{sym}} \tag{8.1}$$

The SEP (or BEP) calculated using (8.1) is in perfect conjunction with the relative frequency approach of probability that states

$$P_e = \lim_{Count_{sym} \to \infty} \frac{Count_{err}}{Count_{sym}} \tag{8.2}$$

Usually the number of symbols (or bits) examined at a SNR point is 10^3 times higher than the inverse of the expected error rate, i.e. to test a SEP (or BEP) of 10^{-4}, 10^7 symbols (or bits) were examined to ensure a confidence band of 99% [3, Fig. 13.42]. Further, an average of 30 individual runs was taken to smooth any variation about the mean.

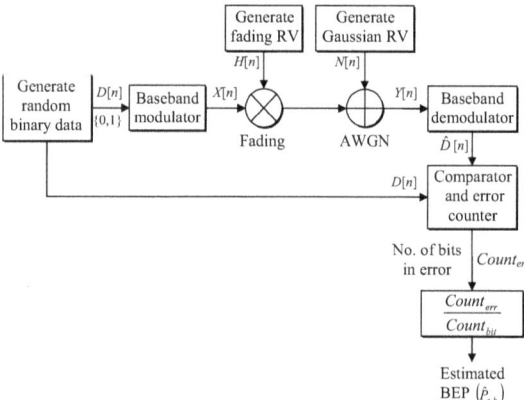

Figure 8.1: Monte Carlo simulation model for estimating BEP of digital signals in fading channels.

The various component blocks of the simulation model realized through MATLAB are described in Fig. 8.1. The subsystems are connected together such that the output of the former block forms the input to the later block. The first block generates a random sequence $D[n]$ of binary digits $\{0, 1\}$ with equal probability. Next the generated bits are modulated through a baseband modulator. Since simulation is based on discrete samples of underlying continuous signals, it is usual for passband systems to be simulated as equivalent baseband processes. The modulated signal $X[n]$ is then fed to the channel where both the system imperfections, fading, and AWGN, are introduced. Accordingly, the simulation process involves generation of two different kinds of RVs, $H[n]$ and $N[n]$. Gaussian noise may be realized rather easily with an in-built function `randn()` available in MATLAB. However, realization of a fading envelope involves specific algorithms for different fading models, and described in Section 8.3. If the receiver consists of a diversity combiner, several such noise and fading samples need to be generated. The simulation of diversity combiner is discussed separately in Section 8.4. The output of the channel (or combiner), denoted by $Y[n]$, is passed through a baseband demodulator to reproduce an estimate $\hat{D}[n]$ of the original signal. The detected sequence at the demodulator output is compared bit by bit with the (error free) transmitted sequence and errors are counted. The estimated BEP is given by

$$\hat{P}_{e,b} = \frac{Count_{err}}{Count_{bit}} \tag{8.3}$$

For SEP estimation, the same model may be used with an additional block, a bit to symbol mapper, placed before the modulator. The demodulator, in that case, would give out demodulated symbols, which may be compared with the original outputs from the mapper block with the help of a comparator. The final SEP value is calculated using (8.1).

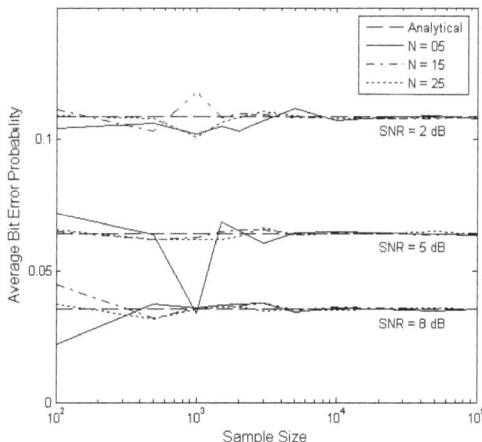

Figure 8.2: Simulated BEP versus sample size.

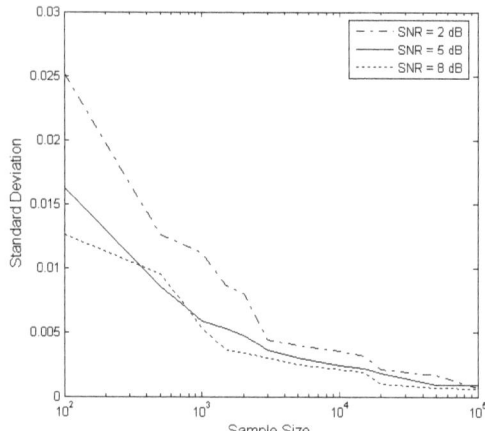

Figure 8.3: Standard deviation versus sample size for $N = 20$.

SNR = 2 dB (BEP = 0.1085)			SNR = 5 dB (BEP = 0.0642)			SNR = 8 dB (BEP = 0.0355)		
Sample size	Avg. BEP	Std. dev.	Sample size	Avg. BEP	Std. dev.	Sample size	Avg. BEP	Std. dev.
500	0.1047	0.0126	500	0.0662	0.0086	500	0.0356	0.0096
1000	0.1076	0.0112	1000	0.0654	0.0059	1000	0.0350	0.0054
1500	0.1089	0.0087	1500	0.0621	0.0053	1500	0.0345	0.0036
2000	0.1065	0.0081	2000	0.0634	0.0048	2000	0.0363	0.0034
3000	0.1083	0.0044	3000	0.0626	0.0036	3000	0.0362	0.0030
5000	0.1091	0.0040	5000	0.0659	0.0030	5000	0.0359	0.0025
10000	0.1095	0.0035	10000	0.0642	0.0024	10000	0.0356	0.0021
15000	0.1082	0.0032	15000	0.0649	0.0022	15000	0.0354	0.0019
20000	0.1085	0.0021	20000	0.0638	0.0018	20000	0.0349	0.0010
50000	0.1086	0.0017	50000	0.0643	0.0009	50000	0.0356	0.0007

Table 8.1: Variation of average BEP and standard deviation with sample size.

The model described above is simulated with MATLAB for BPSK modulation and Rayleigh fading channel. BEP has been calculated as a function of sample size (length of signal) taking average of $N(= 5, 15,$ and 25) individual runs. Effect of sample size on the simulation result has been investigated for SNR values of 2 dB, 5 dB and 8 dB. The results alongwith corresponding theoretical values (dashed lines) have been plotted in Fig. 8.2. In general, average of all replications approach theoretical values more rapidly than individual replications. Therefore, the results improve when N increases. Also, simulation results exhibit close fit as sample size increases. Both the observations lead to a conclusion that better reliability can be achieved at the cost of higher computational complexity.

Standard deviations have been calculated for each sample size using the following equation

$$\text{standard deviation} = \sqrt{\frac{\sum_{i=1}^{N}\left[\hat{P}_{e,b}(i) - \hat{P}_{e,b_{avg}}\right]^2}{N}} \qquad (8.4)$$

where, $\hat{P}_{e,b}(i)$ denotes the estimated BEP obtained for ith run and $\hat{P}_{e,b_{avg}}$ is the average BEP of all the runs. Average standard deviation of $N(=20)$ runs is taken for each SNR value and plotted in Fig. 8.3 with respect to the sample size, and using a logarithmic scale for the horizontal axis. The accuracy of the mean value of all N replications increases exponentially with sample size, as evident by the decrease of the standard deviation.

The detailed result is presented in Table 8.1 from which one can clearly identify how standard deviation can be reduced by increasing the signal length. The simulation results show negligible improvement in the higher range of sample size, and point toward diminishing returns when sample size is increased. An appropriate choice of sample size is thus important, and in practice, is determined by the modulation, shaping, coding, and filtering constraints.

8.2 Simulation of Digital Signals

8.2.1 Binary Modulations

Random binary data $D[n]$ that follows a uniform distribution may be generated with MATLAB as

```
D=randint(1,num_bit);
```

which is then modulated with baseband modulators. For binary coherent modulations, the baseband modulators behave as simple mappers and their implementation requires a single line of code. For BPSK, the mapping rule is, $\{0 \rightarrow -1, 1 \rightarrow +1\}$ and the modulator may be realized as

```
X=2*data-1;
```

while for BFSK, the mapping is $\{0 \rightarrow j, 1 \rightarrow 1\}$, and the corresponding MATLAB code is

```
X=D+j*(~ D);
```

The demodulation and error checking, for both coherent BPSK and BFSK, may be combined together to reduce coding exercise. If `Error` denotes the error counter, the total error count is obtained as

```
for n=1:num_bit
    if ((Y(n)>0 && D(n)==0)||(Y(n)<0 && D(n)==1))
        Error=Error+1;
    end
end
```

for BPSK, and

```
Z=real(Y)-imag(Y);
for n=1:num_bit
    if ((Z(n)>0 && D(n)==0)||(Z(n)<0 && D(n)==1))
        Error=Error+1;
    end
end
```

for BFSK. The `for` loop is used for taking decisions bit-by-bit and incrementing the error counter when a mismatch is found. In case of BFSK, it may be noted that the decision variable `Z` has to be computed before taking the decision. Further, for all modulations, the BEP can be calculated by dividing the total number of errors (`Error`) by the total bit count (`num_bit`)

```
Error=Error/num_bit;
```

For non-coherent and differentially coherent schemes, however, the baseband simulation approach does not work well. A passband simulation is thus preferred for BEP simulation of DPSK and ncBFSK. Accordingly, we define the bit period, sampling rate, and basis functions in the preamble

```
T=1;                              %Bit period
os=10;                            %Sampling rate
phi1=cos(2*n_f*pi*[0:T/os:T-T/os]);  %Basis function #1
phi2=sin(2*n_f*pi*[0:T/os:T-T/os]);  %Basis function #2
```

The parameter `n_f` is an integer denoting the ratio f_c/f_b, where f_c is the carrier frequency, and $f_b = 1/T$. Although the ratio is quite high for practical systems, a small value of $n\_f(< 10)$ is enough for the simulation purpose.

In case of passband DPSK, the modulation involves both differential encoding and frequency translation. The first operation is realized through xor function, while for the second, we multiply the baseband data with the in-phase basis function phi1

```
for n=2:num_bit    %XOR operation
    B(n)=xor(B(n-1),D(n));
end
S=2*B-1;            %NRZ level encoding
for n=1:num_bit    %Passband modulation
    X(1+(n-1)*os:n*os)=S(n).*phi1(1:os);
end
```

MATLAB code for the demodulator may be developed following the block diagram [1, Fig. 3.5.3(a)] as

```
for n=1:num_bit    %Demodulation
    I(n)=Y(1+(n-1)*os:n*os)*phi1'/os;
    Q(n)=Y(1+(n-1)*os:n*os)*phi2'/os;
end
for n=2:num_bit    %Error counting
Z(n)=I(n)*I(n-1)+Q(n)*Q(n-1);
    if ((Z(n)>0 && D(n)==1)||(Z(n)<0 && D(n)==0))
        Error=Error+1;
    end
end
Error=Error/(num_bit-1);
```

where the error probability is estimated based on num_bit-1 bits. The code for ncBFSK modulation and demodulation are not discussed separately in this section. It is a special case of ncMFSK for $L = 2$, and the general coding is described in Section 8.2.3.

8.2.2 MPSK

For general M-ary PSK, the baseband modulator may be designed as

```
D=randint(1,num_sym,M);
phi=(2*(D+1)-1)*pi/M;
X=exp(j*phi);
```

205

to produce output symbols that follow a constellation diagram given in Fig. 2.6. For demodulation and SEP calculation, the following code may be used

```
for n=1:num_sym
    phir(n)=mod(atan2(imag(Y(n)),real(Y(n))),2*pi);
    for t=1:M
        if ((phir(n)>=2*(t-1)*pi/M)&&(phir(n)<2*t*pi/M))
            Z(n)=t-1;
        end
    end
    if (D(n)~=Z(n))
        Error=Error+1;
    end
end
Error=Error/num_sym;
```

where $Z[n]$ is the reconstructed symbol sequence produced by observing the received phase phir with the help of atan2 function in MATLAB, giving a four-quadrant inverse tangent.

In the present text, however, we have concentrated on the differential M-ary PSK or MDPSK more, and particularly investigated the $\pi/4$-DQPSK in detail. Moreover, the BEP of $\pi/4$-DQPSK, not the SEP, were derived during the analytical calculations. The modulator is simulated through the MATLAB code

```
D=randint(1,num_bit);
phi=0;

for n=1:num_sym
    sign=1-2*D(2*n-1); %First bit 0, sign→ +ve,
                       %First bit 1, sign→ -ve
    val=1+2*D(2*n);    %Second bit 0, val→ 1,
                       %Second bit 1, val→ 3
    dphi(n)=mod(sign*val*(pi/4),2*pi);
    phase=phi+dphi(n),
    X(n)=exp(j*phase);
    phi=phase;
end
```

where num_sym $=$ num_bit$/2$, as every symbol is formed by grouping two successive bits. The two variables sign and val are used to find the incremental phase difference dphin following the mapping rule given by Table 2.1. The code for $\pi/4$-DQPSK demodulator is

206

given below

```
for n=2:num_sym
    I(n)=real(Y(n))*real(Y(n-1))+imag(Y(n))*imag(Y(n-1));
    Q(n)=imag(Y(n))*real(Y(n-1))-real(Y(n))*imag(Y(n-1));
    if (Q(n)>0)  %Decision for the first bit
        Z(2*n-1)=0;
    else
        Z(2*n-1)=1;
    if (I(n)>0)  %Decision for the second bit
        Z(2*n)=0;
    else
        Z(2*n)=1;
end
```

which was developed following an equivalent complex baseband implementation of the receiver structure presented in Fig. 2.10. The BEP calculation follows after reconstruction of the bit sequence $Z[n]$

```
for n=3:num_bit
if (D(n)~=Z(n))
        Error=Error+1;
    end
end
Error=Error/(num_bit-2);
```

where, like the DPSK case, we have omitted the first symbol (or the first two bits).

8.2.3 MFSK

For MFSK modulation the equivalent baseband signal is given by (2.100) where $i = 1, 2, \cdots, M$. The approach we took so far, i.e. to use one sample per bit/ symbol of the baseband waveform is no longer valid, as we have to realize M orthogonal baseband signals. For simulating MFSK, we sample the baseband signal itself. The number of samples (os) should be at least $M + 1$ or more in order to maintain orthogonality.

The baseband modulator code is as follows

```
D=randint(1,num_sym,M);
for n=1:num_sym
    X((n-1)*os+1:n*os)=exp(j*pi*(D(n)+1)*[0:T/os:T-T/os]);
end
```

which may be used for both coherent and non-coherent demodulation. Of course, the demodulator code is different for the two cases and given by

```
for n=1:num_sym
    Ydref=0; index=0;
    for m=1:M
        Yd=Y((n-1)*os+1:n*os)*exp(j*pi*m*[0:T/os:T-T/os])'/os;
        if (real(Yd)>Ydref)
            Ydref=real(Yd);
            index=m;
        end
    end
    Z(n)=index;
    if (Z(n) =(D(n)+1))
        Error=Error+1;
    end
end
```

for the coherent case, and the following code may be used

```
for n=1:num_sym
    Ydref=0; index=0;
    for m=1:M
        I=real(Y((n-1)*os+1:n*os))*cos(pi*m*[0:T/os:T-T/os])'/os;
        Q=imag(Y((n-1)*os+1:n*os))*sin(pi*m*[0:T/os:T-T/os])'/os;
        Yd=(I^2)+(Q^2);

        if (Yd>Ydref)
            Ydref=Yd;
            index=m;
        end
    end
    Z(n)=index;
    if (Z(n) =(D(n)+1))
        Error=Error+1;
    end
end
```

for the non-coherent case. In both the algorithms Yd serves as the decision variable. To reconstruct the symbol, we compare the decision variable against a reference Ydref, which is initialized to the correlator/ matched filter output of the first branch. The variable

208

index stores the index number of the branch giving maximum Yd value and assigns the decoded symbol sequence $Z[n]$ with the corresponding symbol from the alphabet. The inner for loop is used for searching maximum value of Yd.

8.2.4 MQAM

After discussing simulation of binary, MPSK, and MFSK modulation, in this section, we would discuss SEP simulation of square M-ary QAM. A modulation order of $M = 2^4 = 16$ is chosen for the code described here. Naturally, num_sym = num_bit/4. Also, the bit sequence that is fed to the modulator is assumed to be Gray coded. The MATLAB code for a 16QAM modulator is given by

```
D=randint(1,num_bit);
for n=1:num_sym
    g1=2*(D(4*n-3))-1;
    g2=2*(xor(D(4*n-3),(D(4*n-2))))-1;
    XI=2*g1+g2;
    g3=2*(D(4*n-1))-1;
    g4=2*(xor(D(4*n-1),(D(4*n))))-1;
    XQ=2*g3-g4;
    X(n)=XI+j*XQ;
end
```

where the data is converted back to signed binary code (from Gray code) during the modulation. As an example, if the gray coded input data fed to the modulator is 1101 (for 16QAM every symbol contains four bits), then g1 = +1, g2 = −1, g3 = −1, and g4 = +1. This gives a value for the in-phase and quadrature component of the modulated signal as XI = +1 and XQ = +1, and the output baseband symbol is X = 1 + j.

The demodulation requires decisions for each bit separately

```
for n=1:num_sym
    Z(4*n-3)=(sign(real(Y(n)))+1)/2;
    Z(4*n-1)=(-sign(imag(Y(n)))+1)/2;
    if (abs(real(Y(n)))>2)
        Z(4*n-2)=0;
    else
        Z(4*n-2)=1;
    if (abs(imag(Y(n)))>2)
        Z(4*n)=0;
    else
        Z(4*n)=1;
    end
end
```

In the decoded bit sequence $Z[n]$, if any bit out of the four that belongs to a symbol is erroneous, a symbol error occurs. Thus, the SEP may be calculated as

```
for n=1:num_sym
    if ((Z(4*n-3)~=D(4*n-3))||((Z(4*n-2)~=D(4*n-2))||...
       (Z(4*n-1)~=D(4*n-1))||((Z(4*n)~=D(4*n)))
        Error=Error+1;
    end
end
```

8.3   Simulation of Fading Envelopes

All the fading models used in the earlier chapters may be simulated by exploiting the relationship between the fading RV and Gaussian RVs. Generation of Gaussian RVs are relatively easy in MATLAB environment as the software package provides an in-built function `randn()` for the purpose. From the theory of RVs, it can be shown that the modulus of a complex Gaussian process $X_1(t) + jX_2(t)$, i.e.

$$R(t) = |X_1(t) + jX_2(t)| = \sqrt{X_1^2(t) + X_2^2(t)} \tag{8.5}$$

where $X_{1,2}(t)$ are uncorrelated low pass Gaussian processes with means $m_{1,2}$ and of variances $\sigma_{1,2}^2$ respectively, follows Rayleigh, Rician, and Hoyt distribution under the following conditions

1. **Case 1:** $m_1 = m_2 = 0, \sigma_1 = \sigma_2$ $\quad \Rightarrow$ Rayleigh distribution

2. **Case 2:** $m_1 = +ve, m_2 = 0, \sigma_1 = \sigma_2 \Rightarrow$ Rician distribution

3. **Case 3:** $m_1 = m_2 = 0, \sigma_1 \neq \sigma_2 \qquad \Rightarrow$ Hoyt distribution

Although, the Nakagami-m random process cannot be expressed as a special case of the random process $R(t)$ like other models given above, it is also closely related with Gaussian process. If $\{X_1, X_2, \cdots, X_{2m}\}$ denotes a set of Gaussian RV with zero mean and equal variance, i.e. $X_i \sim \mathcal{N}(0, \sigma^2); \forall i$, then a RV, R, defined as $R = \sqrt{X_1^2 + X_2^2 + \cdots + X_{2m}^2}$, follows Nakagami distribution.

Algorithm 1 Simulation of Rayleigh fading envelope

1: $\Omega \leftarrow 1$

2: $\sigma \leftarrow \sqrt{\dfrac{\Omega}{2}}$

3: **generate** $\mathbf{H}_I \sim \mathcal{N}(0, \sigma^2)$
4: **generate** $\mathbf{H}_Q \sim \mathcal{N}(0, \sigma^2)$
5: **return** $\mathbf{H} = \sqrt{\mathbf{H}_I^2 + \mathbf{H}_Q^2}$

The detailed algorithms for generating the four kind of fading RVs are given by Algorithm 1 - Algorithm 4. The first line of each algorithm ensures that the average fading power, $E\{\alpha^2\} = \Omega$, is kept fixed at 1. Rest of the algorithm is, however, designed to work with arbitrary Ω values.

Algorithm 2 Simulation of Rician fading envelope

1: $\Omega \leftarrow 1$
2: $K \leftarrow 10^{(K_{\text{dB}}/10)}$

3: $S \leftarrow \sqrt{\dfrac{K\Omega}{1 + K}}$

4: $\sigma \leftarrow \sqrt{\dfrac{\Omega}{2(1 + K)}}$

5: **generate** $\mathbf{H}_I \sim \mathcal{N}(S, \sigma^2)$
6: **generate** $\mathbf{H}_Q \sim \mathcal{N}(0, \sigma^2)$
7: **return** $\mathbf{H} = \sqrt{\mathbf{H}_I^2 + \mathbf{H}_Q^2}$

For Rician fading, the fading parameter K is generally expressed in dB. Thus, in the second line of Algorithm 2, we first convert the parameter to normal scale. Next, the mean S of the in-phase Gaussian process, and standard deviation σ (for in-phase as well as quadrature Gaussian process) are calculated using the expressions derived in Section 3.2.4.

As pointed out earlier, the Nakagami PDF may be represented as the amplitude of a sum of squared independent Gaussian RVs

$$R = \sqrt{X_1^2 + X_2^2 + \cdots + X_n^2} \tag{8.6}$$

Algorithm 3 Simulation of Nakagami fading envelope
___
1: $\Omega \leftarrow 1$
2: $a \leftarrow m$
3: $b \leftarrow \dfrac{\Omega}{m}$
4: **generate** $\mathbf{H}_{sq} \sim \mathcal{G}(a, b)$
5: **return** $\mathbf{H} = \sqrt{\mathbf{H}_{sq}}$
___

where $X_i; i = 1, \cdots, n$ are IID Gaussian RVs, each with zero-mean and of variance σ_X^2. The RV, R, in (8.6), has a Nakagami PDF with parameter $m = n/2$ and second moment $\Omega = 2m\sigma_X^2$. However, one can generate Nakagami RVs with only integer and half-integer values of m through the method, and the simulation becomes computationally inefficient for higher values of m [202]. In Algorithm 3, we have used square root of a gamma RV $\mathcal{G}(a, b)$ with parameters $a = m, b = \Omega/m$ to generate Nakagami RVs for arbitrary m values. Gamma RVs may be generated with MATLAB using the `gamrnd()` function. A simple procedure for generating gamma RVs is also given in [203].

Algorithm 4 Simulation of Hoyt fading envelope
___
1: $\Omega \leftarrow 1$
2: $\sigma_1 \leftarrow \sqrt{\dfrac{\Omega q^2}{1 + q^2}}$
3: $\sigma_2 \leftarrow \sqrt{\dfrac{\Omega}{1 + q^2}}$
4: **generate** $\mathbf{H}_I \sim \mathcal{N}(0, \sigma_1^2)$
5: **generate** $\mathbf{H}_Q \sim \mathcal{N}(0, \sigma_2^2)$
6: **return** $\mathbf{H} = \sqrt{\mathbf{H}_I^2 + \mathbf{H}_Q^2}$
___

The instantaneous amplitude $R(t)$, of Nakagami-q or Hoyt fading process is the modulus of a complex Gaussian process $X_1(t) + jX_2(t)$ [204]

$$R(t) = |X_1(t) + jX_2(t)| = \sqrt{X_1^2(t) + X_2^2(t)} \tag{8.7}$$

where $X_1(t)$ and $X_2(t)$ are uncorrelated zero mean low pass Gaussian processes of variances σ_1^2 and σ_2^2, respectively. The PDF of the Nakagami-q process, $f_R(\alpha)$ is given by [204, 205]

$$f_R(\alpha) = \frac{\alpha}{\sigma_1 \sigma_2} \exp\left[-\frac{\alpha^2}{4}\left(\frac{1}{\sigma_1^2} + \frac{1}{\sigma_2^2}\right)\right] I_0\left[\frac{\alpha^2}{4}\left(\frac{1}{\sigma_2^2} - \frac{1}{\sigma_1^2}\right)\right] \tag{8.8}$$

where α is the instantaneous fading amplitude. Representing (8.8) in terms of the parameters σ_2 and $q = \sigma_1/\sigma_2$ we have

$$f_R(\alpha) = \frac{\alpha}{q\sigma_2^2} \exp\left[-\frac{\alpha^2}{4\sigma_2^2}\left(1 + \frac{1}{q^2}\right)\right] I_0\left[\frac{\alpha^2}{4\sigma_2^2}\left(1 - \frac{1}{q^2}\right)\right] \tag{8.9}$$

Further we define the average fading power as $E\{R^2\} = \bar{\alpha^2} = \Omega$. Naturally,

$$\Omega = \sigma_1^2 + \sigma_2^2 = \sigma_2^2 \left(1 + q^2\right) \tag{8.10}$$

Making the substitution $\sigma_2^2 = \Omega/\left(1 + q^2\right)$ we arrive at

$$f_R(\alpha) = \frac{(1+q^2)\,\alpha}{q\Omega} \exp\left[-\frac{\left(1+q^2\right)^2 \alpha^2}{4q^2\Omega}\right] I_0 \left[\frac{\left(1-q^4\right)\alpha^2}{4q^2\Omega}\right] \tag{8.11}$$

as mentioned in Section 3.2.6. From the two relations $q = \sigma_1/\sigma_2$ and $\Omega = \sigma_1^2 + \sigma_2^2$ we have

$$\sigma_2^2 = \frac{\Omega}{1+q^2} \Rightarrow \sigma_2 = \sqrt{\frac{\Omega}{1+q^2}} \tag{8.12a}$$

$$\sigma_1^2 = \Omega - \sigma_2^2 \Rightarrow \sigma_1 = \sqrt{\frac{\Omega q^2}{1+q^2}} \tag{8.12b}$$

Algorithm 4 presents the basic methodology to generate Hoyt distributed fading envelope samples with the help of these equations.

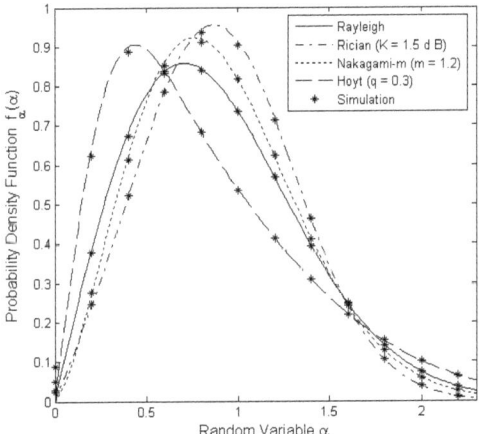

Figure 8.4: Analytical and simulated fading envelopes.

Fig. 8.4 depicts that the simulated PDFs (generated with these algorithms) match exactly with the theoretical PDF expressions.

8.4 Simulation of Diversity Combiners

In the previous chapters, we have investigated four different types of diversity combining, MRC, EGC, SC, and SWC. For all the combining algorithms, it is necessary to first realize L different fading envelopes and AWGN vectors as demonstrated in Fig. 8.5. This can be accomplished by using two dimensional (2D) arrays or cells, where the rows of the generated 2D matrices denote the diversity branch ($i = 1, 2, \cdots, L$) and columns represent individual symbols or bits (indexed by n). The L different inputs to the diversity combiner may be mathematically characterized as

$$Y_i[n] = H_i[n]X[n] + N_i[n] \quad ; i = 1, 2, \cdots, L \tag{8.13}$$

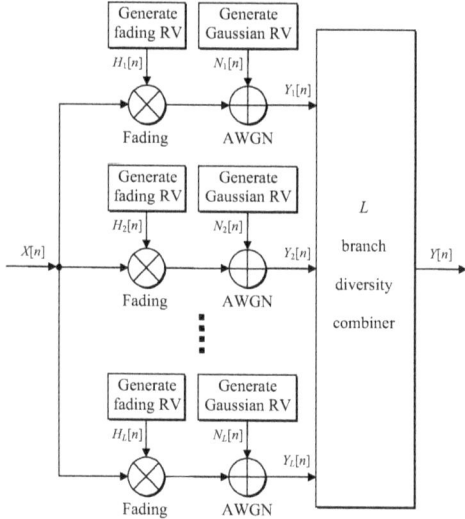

Figure 8.5: Simulation model for multichannel diversity reception.

For MRC, the combining rule is given by the following equation

$$Y[n] = \frac{1}{L} \sum_{i=1}^{L} H_i[n] Y_i[n] \tag{8.14}$$

where each input $Y_i[n]$ to the combiner is weighted by the corresponding fading envelope.

For the EGC case, the rule reduces to

$$Y[n] = \frac{1}{L} \sum_{i=1}^{L} Y_i[n] \tag{8.15}$$

with no such weight vectors. Note that, we have assumed zero phase shift during the transmission of symbols and thus no phase coherer is required at the combiner. Also, for both MRC and EGC, the output is scaled by a factor $1/L$ to keep the average power of the output signal normalized to unity.

In case of SC, the combiner chooses the strongest signal branch as the output, i.e. $Y[n] = Y_j[n]$ where $|Y_j[n]| = \max(\{|Y_i[n]|\}_{i=1}^{L})$. The same is implemented through the following MATLAB code

```
for n=1:num_sym
    temp1=abs(Y{1}(n));
    temp2=Y{1}(n);

    for m=2:L
        if (temp1<abs(Y{m}(n)))
            temp1=abs(Y{m}(n));
            temp2=Y{m}(n);
        end
    end
    Ysc(n)=temp2;
end
```

where the outer for loop runs the simulation for each symbol to be tested. The two temporary variables, temp1 and temp2, keep track of the absolute and signed values of the signal envelope, and initially store the absolute and signed values of the signal in branch 1. The inner for loop tests all the input branches, and reassigns temp1 and temp2 with values from branch j if $|Y_j[n]| > |Y_1[n]|$. The final output of the combiner is stored in Ysc.

As SWC is a suboptimal version of SC, its implementation through MATLAB results in a code that is somewhat similar to the previous code described in this section. There are two variants of SWC, SSC and SEC. However, our discussions are restricted to the general SEC case only because SSC is a special case of SEC for $L = 2$. The following lines of code were used to realize an SEC diversity receiver block,

```
for n=1:num_sym
    for m=1:L
        instSNR{m}(n)=((H{m}(n))^2)*SNR;
    end
    if(instSNR{token}(n)<Th)
        for m=1:L
            if(instSNR{m}(n)>Th)
                token=m;
            end
        end
    end
    Ysec(n)=Y{token}(n);
end
```

where the first inner **for** loop is for calculating the instantaneous SNR values, $\gamma_i = \alpha_i^2(E_s/N_0); i = 1, 2, \cdots, L$, of each input for a given symbol. If the current branch (indexed by the variable **token**) SNR is below the specified threshold SNR (**Th**), the switching action is initiated. The second inner **for** loop examines each branch, and selects the first one having a SNR above the threshold. Variable **token** is also modified accordingly. The final output of the combiner is given by **Ysec**.

Chapter Summary

This final chapter discusses the simulation methodology used for verifying different analytical expressions obtained throughout the book. Simulation programs were written in MATLAB as it is widely available and highly capable of matrix manipulations. However, we have deliberately avoided use of any specialized built-in routines and thus these programs can be developed easily with some other programming environment.

The chapter starts with the basic concept of Monte Carlo simulation which is explained with a block diagram and different aspects of the technique has been illustrated with a simple example concerning BPSK. In the next section, codes for simulating modulator and demodulator structures for various digital modulation schemes are described. Simple algorithms for generating fading environments are given next. Finally, the idea of realizing parallel fading channels and combining those channels with different diversity combiners are explained.

Appendix A

Integrals of Marcum's Q Function

The following integral after Simon et al. [73, (5.56)]

$$\mathcal{I}_1 = \int_0^\infty x^{m-1} \exp(-p^2 x/2) Q_j(a\sqrt{x}, b\sqrt{x}) dx \tag{A.1}$$

can be written as

$$
\begin{aligned}
\mathcal{I}_1 = \ & \left(\frac{2}{p^2}\right)^m (m-1)! \\
& \times \left[1 + \left(\frac{b^2}{s}\right)^j \sum_{i=0}^{m-1} \binom{i+j}{i} \left(\frac{p^2}{s}\right)^i \right. \\
& \times \left\{ \frac{a^2}{s} \, {}_2F_1\left(\frac{i+j+1}{2}, \frac{i+j+2}{2}; j+1; \frac{4a^2b^2}{s^2}\right) \right. \\
& \left. \left. - \frac{j}{i+j} \, {}_2F_1\left(\frac{i+j}{2}, \frac{i+j+1}{2}; j; \frac{4a^2b^2}{s^2}\right) \right\} \right]
\end{aligned}
\tag{A.2}
$$

where $\{m, j\}$ are arbitrary integers independent of each other, $s = p^2 + a^2 + b^2$. A similar kind of integral may also be found in Okui's paper [175].

Specializing for $j = 1$ and letting $p^2/2 = t$, $m - 1 = n$, and $\{a, b\} = \sqrt{2 \pm \sqrt{2}}$, the integral becomes

$$
\begin{aligned}
\mathcal{I}_2 = \ & \frac{n!}{t^{n+1}} \left[1 + \frac{1}{2(t+2)^2} \sum_{i=0}^n (i+1) \left(\frac{t}{t+2}\right)^i \right. \\
& \times {}_2F_1\left(\frac{i+2}{2}, \frac{i+3}{2}; 2; \frac{2}{(t+2)^2}\right) - \left(\frac{2 \mp \sqrt{2}}{2t+4}\right) \\
& \left. \times \sum_{i=0}^n \left(\frac{t}{t+2}\right)^i {}_2F_1\left(\frac{i+1}{2}, \frac{i+2}{2}; 1; \frac{2}{(t+2)^2}\right) \right]
\end{aligned}
\tag{A.3}
$$

From (A.3) it can be shown

$$\int_0^\infty x^n \exp(-tx) \left[Q\left(\sqrt{\alpha x}, \sqrt{\beta x}\right) - Q\left(\sqrt{\beta x}, \sqrt{\alpha x}\right) \right] dx$$

$$= \sqrt{2}n! \sum_{i=0}^n \frac{t^{i-n-1}}{(t+2)^{i+1}} \, {}_2F_1\left(\frac{i+1}{2}, \frac{i+2}{2}; 1; \frac{2}{(t+2)^2}\right) \tag{A.4}$$

by noting that some of the terms vanish due to the symmetrical nature of the integral. In (A.4), $\{\alpha, \beta\} = 2 \pm \sqrt{2}$ and the subscript of Marcum's Q function has been removed for minimizing notations. Further when $n = 0$, from (A.4) and by applying [8, (9.121.1)] we have

$$\int_0^\infty \exp(-tx) \left[Q\left(\sqrt{\alpha x}, \sqrt{\beta x}\right) - Q\left(\sqrt{\beta x}, \sqrt{\alpha x}\right) \right] dx$$

$$= \frac{1}{t} \sqrt{\frac{2}{t^2 + 4t + 2}} \tag{A.5}$$

Appendix B

Integrals of Modified Bessel Function

B.1 Integrations Involving $I_v(bx)$, x^n, and $\exp(-mx)$

From [8, (6.622.3)]

$$\int_0^\infty x^{\mu-1} \exp(-x\cosh\alpha)I_v(x)dx$$

$$= \sqrt{\frac{2}{\pi}} \exp\left[-\frac{j(2\mu-1)\pi}{2}\right] \frac{Q_{v-1/2}^{\mu-1/2}(\cosh\alpha)}{(\sinh\alpha)^{\mu-1/2}}$$

(B.1)

where $\Re\{\mu+v\} > 0$, $\Re\{\cosh\alpha\} > 1$, $j = \sqrt{-1}$, and $Q_v^\mu(\cdot)$ is the associated Legendre function of the second kind and defined as [8, (8.703)]

$$Q_v^\mu(z) = \exp(j\mu\pi)\frac{\Gamma(\mu+v+1)\Gamma(1/2)}{2^{v+1}\Gamma(v+3/2)}\frac{(z^2-1)^{\mu/2}}{z^{\mu+v+1}}$$

$$\times {}_2F_1\left(\frac{\mu+v+2}{2}, \frac{\mu+v+1}{2}, v+\frac{3}{2}; \frac{1}{z^2}\right)$$

(B.2)

Letting $\cosh\alpha = t$ and using the definition of $Q_v^\mu(\cdot)$ we have

$$\int_0^\infty x^{\mu-1}\exp(-tx)I_v(x)dx$$

$$= \frac{1}{2^v t^{\mu+v}}\frac{\Gamma(\mu+v)}{\Gamma(v+1)} {}_2F_1\left(\frac{\mu+v+1}{2}, \frac{\mu+v}{2}; v+1; \frac{1}{t^2}\right)$$

(B.3)

When $\mu = 1$, using the identity [9, (15.1.14)]

$$\int_0^\infty \exp(-tx)I_v(x)dx = \frac{1}{\sqrt{t^2-1}}\left(t - \frac{1}{\sqrt{t^2-1}}\right)^v$$

(B.4)

Further, as

$$\int_0^\infty x^{\mu-1}\exp(-tx)I_\nu(\beta x)dx = \frac{1}{\beta^\mu}\int_0^\infty z^{\mu-1}\exp\left(-z\frac{t}{\beta}\right)I_\nu(z)dz \tag{B.5}$$

we may generalize the result in (B.3)

$$\int_0^\infty x^{\mu-1}\exp(-tx)I_\nu(\beta x)dx$$

$$= \frac{\beta^\nu}{2^\nu t^{\mu+\nu}}\frac{\Gamma(\mu+\nu)}{\Gamma(\nu+1)}\,_2F_1\left(\frac{\mu+\nu+1}{2},\frac{\mu+\nu}{2};\nu+1;\frac{\beta^2}{t^2}\right) \tag{B.6}$$

For $\mu = 1$ and $\nu = 0$, from (B.6) and using the identity [9, (15.1.8)]

$$\int_0^\infty \exp(-tx)I_0(\beta x)dx = \left[t^2-\beta^2\right]^{-1/2} \tag{B.7}$$

which is nothing but a special case of [8, (6.611.4)], namely

$$\int_0^\infty \exp(-tx)I_\nu(\beta x)dx = \frac{1}{\sqrt{t^2-\beta^2}}\left[\frac{t-\sqrt{t^2-\beta^2}}{\beta}\right]^\nu \tag{B.8}$$

B.2   Integrations Involving $I_v(b\sqrt{x})$, x^n, and $\exp(-mx)$

From [8, (6.643.2)]

$$\int_0^\infty x^{\mu-1/2}\exp(-\alpha x)I_{2\nu}(2\beta\sqrt{x})dx$$

$$= \frac{\Gamma(\mu+\nu+1/2)}{\beta\alpha^\mu\Gamma(2\nu+1)}\exp\left(\frac{\beta^2}{2\alpha}\right)M_{-\mu,\nu}\left(\frac{\beta^2}{\alpha}\right) \tag{B.9}$$

where $\Re\{\mu+\nu+1/2\} > 0$ and $M_{\mu,\nu}(\cdot)$ is the Whittaker's function of the first kind and defined as [8, (9.220.2)]

$$M_{\mu,\nu}(z) = z^{\nu+1/2}\exp\left(-\frac{z}{2}\right)\,_1F_1\left(\nu-\mu+\frac{1}{2};2\nu+1;z\right) \tag{B.10}$$

From (B.9) and (B.10)

$$\int_0^\infty x^{\mu-1/2}\exp(-\alpha x)I_\nu(\beta\sqrt{x})dx$$

$$= \frac{2\Gamma(\mu+\frac{\nu+1}{2})}{\beta\alpha^\mu\Gamma(\nu+1)}\left(\frac{\beta^2}{4\alpha}\right)^{\frac{\nu+1}{2}}\,_1F_1\left(\mu+\frac{\nu+1}{2};\nu+1;\frac{\beta^2}{4\alpha}\right) \tag{B.11}$$

For the special case, $\mu = 1/2$, $\nu = 0$, the reduced result is

$$\int_0^\infty \exp(-\alpha x) I_0(\beta\sqrt{x}) dx = \frac{1}{\alpha} \exp\left(\frac{\beta^2}{4\alpha}\right) \tag{B.12}$$

which was obtained using [8, (9.215.1)].

B.3 Integrations Involving $K_v(b\sqrt{x})$, x^n, and $\exp(-mx)$

From [133, (2.16.8.4)]

$$\int_0^\infty x^{\alpha-1} \exp(-\beta x^2) K_\nu(tx) dx$$

$$= \frac{\beta^{\frac{1-\alpha}{2}}}{2t} \Gamma\left(\frac{\alpha+\nu}{2}\right) \Gamma\left(\frac{\alpha-\nu}{2}\right) \exp\left(\frac{t^2}{8\beta}\right) W_{\frac{1-\alpha}{2},\frac{\nu}{2}}\left(\frac{t^2}{4\beta}\right) \tag{B.13}$$

where $\Re\{\alpha\} > \Re|v|$, $\Re\{\beta\} > 0$ and $W_{\mu,\nu}(\cdot)$ is the Whittaker's function of the second kind and bearing a close relation [9, (13.1.33)] with Tricomi's confluent hypergeometric function

$$W_{\mu,\nu}(z) = z^{\nu+1/2} \exp\left(-\frac{z}{2}\right) U\left(\nu - \mu + \frac{1}{2}; 2\nu + 1; z\right) \tag{B.14}$$

Now, combining (B.13) and (B.14) with a substitution $x = \sqrt{z}$ and writing α instead of $\alpha/2$, we finally have

$$\int_0^\infty z^{\alpha-1} \exp(-\beta z) K_\nu(t\sqrt{z}) dz$$

$$= \frac{1}{2\beta^\alpha} \left(\frac{t^2}{4\beta}\right)^{\frac{\nu}{2}} \Gamma\left(\alpha + \frac{\nu}{2}\right) \Gamma\left(\alpha - \frac{\nu}{2}\right) U\left(\alpha + \frac{\nu}{2}; \nu + 1; \frac{t^2}{4\beta}\right) \tag{B.15}$$

Bibliography

[1] S. G. Wilson. *Digital Modulation and Coding*. Prentice Hall, NJ, USA, 1996.

[2] J. G. Proakis. *Digital Communications*. McGraw-Hill, NY, USA, 4th edition, Aug. 2000.

[3] I. A. Glover and P. M. Grant. *Digital Communications*. Pearson Education, New Delhi, India, 2nd edition, 2008.

[4] A. J. Goldsmith. *Wireless Communications*. Cambridge University Press, NY, USA, Aug. 2005.

[5] J. M. Wozencraft and I. M. Jacobs. *Principles of Communication Engineering*. John Wiley and Sons, NJ, USA, 1965.

[6] F. Xiong. *Digital Modulation Techniques*. Artech House, Norwood, MA, USA, 2nd edition, Apr. 2006.

[7] A. H. Wojner. Unknown bounds on performance in Nakagami channels. *IEEE Trans. Commun.*, 34(1):22–24, Jan. 1986.

[8] I. S. Gradshteyn and I. M. Ryzhik. *Table of Integrals, Series and Products*. Academic Press/ Elsevier, San Diego, CA, USA, 7th edition, Mar. 2007.

[9] M. Abramowitz and I. A. Stegun. *Handbook of Mathematical Functions with Formulas, Graphs, and Mathematical Tables*. Dover, New York, USA, 9th edition, 1970.

[10] J. W. Craig. A new, simple and exact result for calculating the probability of error for two-dimensional signal constellations. In *Proc. IEEE MILCOM*, pages 571–575, Nov. 1991.

[11] M. K. Simon. Single integral representations of certain integer powers of the Gaussian Q-function and their application. *IEEE Commun. Lett.*, 6(12):532–534, Dec. 2002.

[12] L. Zhang, Z. Cao, and C. Gao. Application of RS-coded MPSK modulation scenarios to compressed image communication in mobile fading channel. In *Proc. IEEE VTC*, volume 3, pages 1198–1203, Sep. 2000.

[13] T. T. Ha. *Digital Satellite Communications*. McGraw-Hill, Singapore, 2nd edition, Mar. 1990.

[14] W. P. Osborne, T. J. Wolcott, and J. M. Gardner. A method for evaluating MPSK performance using a $(M/2)$PSK signal set. In *Proc. IEEE 14th Annual Intl. Phoenix Conf. Comp. Commun.*, pages 617–621, Mar. 1995.

[15] A. J. Viterbi and A. M. Viterbi. Nonlinear estimation of PSK-modulated carrier phase with application to burst digital transmission. *IEEE Trans. Info. Th.*, 29(4):543–551, Jul. 1983.

[16] J. Hamkins. SAT02-2: Modulation classification of MPSK for space applications. In *Proc. IEEE GLOBECOMM*, pages 1–5, Nov. 2006.

[17] C. E. Saavedra, M. J. Vaughan, and R. C. Compton. An M-PSK modulator for quasi-optical wireless array applications. In *Proc. IEEE MTT-S Intl. Micro. Symp. Dig.*, volume 3, pages 1243–1246, Jun. 1996.

[18] A. Papoulis and S. U. Pillai. *Probability, Random Variables and Stochastic Processes*. McGraw-Hill, New Delhi, India, 4th edition, 2002.

[19] P. Z. Peebles. *Probability, Random Variables, and Random Signal Principles*. McGraw-Hill, NY, USA, 4th edition, Jul. 2000.

[20] S. Stein and J J. Jones. *Modern Communication Principles*. McGraw-Hill, NY, USA, 1967.

[21] J. Sun and I. S. Reed. Performance of MDPSK, MPSK, and noncoherent MFSK in wireless Rician fading channels. *IEEE Trans. Commun.*, 47(6):813–816, Jun. 1999.

[22] R. F. Pawula. A new formula for MDPSK symbol error probability. *IEEE Commun. Lett.*, 2(10):271–272, Oct. 1998.

[23] S. Sheng. Mobile television receivers: A free-to-air overview. *IEEE Commun. Magz.*, 47(9):142–149, Sep. 2009.

[24] M. K. Simon, S. M. Hinedi, and W. C. Lindsey. *Digital Communication Techniques: Signal Design and Detection*. Prentice Hall, New Delhi, India, 2004.

[25] R. F. Pawula, S. O. Rice, and J. H. Roberts. Distribution of the phase angle between two vectors perturbed by Gaussian noise. *IEEE Trans. Commun.*, 30(8):1828–1841, Aug. 1982.

[26] E. Arthurs and H. Dym. On the optimum detection of digital signals in the presence of white Gaussian noise - A geometric interpretation and a study of three basic data transmission systems. *IRE Trans. Commun.*, 10(4):336–372, Dec. 1962.

[27] T. S. Rappaport. *Wireless Communications: Principles and Practice*. Prentice Hall, New Delhi, 2nd edition, 2003.

[28] H. Schulze and C. Lüders. *Theory and Applications of OFDM and CDMA*. John Wiley and Sons, West Sussex, England, 2005.

[29] IEEE 802.15.3. IEEE Standard for Wireless Personal Area Networks, pages 242–258, 2003.

[30] F. Xiong and D. Wu. Multiple-symbol differential detection of π/4-DQPSK in land mobile satellite communication channels. *IEE Proc. Commun.*, 147(3):163–168, Jun. 2000.

[31] P. Tarasak and V. K. Bhargava. π/4-shifted differential QPSK space-time block codes. *Electron. Lett.*, 39(18):1327–1329, Sep. 2003.

[32] H. G. Balta, M. Kovaci, and M. M. Nafornita. A study on turbo coding systems with π/4 shifted DQPSK modulation. In *Proc. IEEE ISSCS*, volume 1, pages 367–370, Jul. 2005.

[33] O. B. S. Ali, B. L. Agba, C. Cardinal, and F. Gagnon. Performances simulation of flat fading channels in mobile ad hoc networks. In *Proc. URSI ISSSE*, pages 189–191, Jul. 2007.

[34] J. W. Mark and W. Zhuang. *Wireless Communications and Networking*. Prentice Hall, New Delhi, India, 2005.

[35] L. E. Miller and J. S. Lee. BER expressions for differentially detected π/4 DQPSK modulation. *IEEE Trans. Commun.*, 46(1):71–81, Jan. 1998.

[36] J. I. Marcum. A statistical theory of target detection by pulsed radar. Air force project RAND research memorandum RM-754, Rand Corporation, Santa Monica, CA, USA, pages 159–160, Dec. 1947.

[37] M. K. Simon. A new twist on the Marcum Q function and its application. *IEEE Commun. Lett.*, 2(2):39–41, Feb. 1998.

[38] C. Tellambura and V. K. Bhargava. Unified error analysis of DQPSK in fading channels. *Electron. Lett.*, 30(25):2110–2111, Dec. 1994.

[39] M. K. Simon and M. -S. Alouini. A unified approach to the probability of error for noncoherent and differentially coherent modulations over generalized fading channels. *IEEE Trans. Commun.*, 46(12):1625–1638, Dec. 1998.

[40] M. C. Gursoy. On the energy efficiency of orthogonal signaling. In *Proc. IEEE ISIT*, pages 599–603, Jul. 2008.

[41] H. A. Bustamente and R. S. Davies. Data transmission systems for deep space probes. Report AIAA-1968-1103, 5th annual meeting and technical display, American Institute of Aeronautics and Astronautics, Philadelphia, USA, Oct. 1968.

[42] S. Hussain and S. K. Barton. Comparison of coherent and non-coherent modulation schemes for low-data rate hand-held satellite terminals subject to high phase noise. In *Proc. 10th International Conference on Digital Satellite Communications*, volume 1, pages 339–342, May 1995.

[43] R. Walton and M. Wallace. Near maximum likelihood demodulation for M-ary orthogonal signaling. In *Proc. IEEE VTC*, pages 5–8, May 1993.

[44] J. -I. Kim, S. -H. Yoon, S. -J. Kang, and C. -E. Kang. Interference cancellation technique using channel parameter estimation in DS/CDMA system with M-ary orthogonal modulation. *Electron. Lett.*, 34(12):1194–1195, Jun. 1998.

[45] M. C. Gursoy, H. V. Poor, and S. Verdú. On-off frequency-shift keying for wideband fading channels. *EURASIP J. Wireless Commun. Net.*, 2006:Article ID 98564, 1–15, Jun. 2006.

[46] S. Moon and K. Kim. Performance of satellite communication system with FH-MFSK under various jamming environments. In *Proc. IEEE MILCOM*, volume 1, pages 659–663, Oct. 2001.

[47] M. A. Maggenti, T. T. Ha, and T. Pratt. Spread-spectrum multiple access using wideband noncoherent MFSK. *IEEE Trans. Aero. El. Sys.*, 23(6):767–775, Nov. 1987.

[48] G. E. Atkin and H. P. Corrales. A bandwidth efficient signaling for MFSK systems. In *Proc. IEEE VTC*, pages 596–603, Jun. 1988.

[49] R. Sinha and R. D. Yates. An OFDM based multicarrier MFSK system. In *Proc. IEEE VTC*, volume 1, pages 257–264, Sep. 2000.

[50] S. R. Poram. *Performance of M-ary Modulation Schemes over Fading Channels*. ME Thesis, ECE Department, NIT Durgapur, WB, India, Apr. 2008.

[51] L. W. Hughes. A simple upper bound on the error probability for orthogonal signals in white noise. *IEEE Trans. Commun.*, 40(4):670, Apr. 1992.

[52] M. K. Simon and M. -S. Alouini. Probability of error for noncoherent M-ary orthogonal FSK with postdetection switched combining. *IEEE Trans. Commun.*, 51(9):1456–1462, Sep. 2003.

[53] Q. T. Zhang. Error performance of noncoherent MFSK with L-diversity on correlated fading channels. *IEEE Trans. Wireless Commun.*, 1(3):531–539, Jul. 2002.

[54] P. J. Crepeau. Uncoded and coded performance of MFSK and DPSK in Nakagami fading channels. *IEEE Trans. Commun.*, 40(3):487–493, Mar. 1992.

[55] L. Cao, M. Tao, and P. Y. Kam. Closed-form performance of MFSK signals with diversity reception over non-identical fading channels. In *Proc. IEEE WCNC*, pages 740–745, Mar. 2007.

[56] R. M. Radaydeh and M. M. Matalgah. Compact formulas for the average error performance of noncoherent M-ary orthogonal signals over generalized Rican, Nakagami-m, and Nakagami-q fading channels with diversity reception. *IEEE Trans. Commun.*, 56(1):32–38, Jan. 2008.

[57] A. Annamalai and C. Tellambura. A moment-generating function (MGF) derivative-based unified analysis of incoherent diversity reception of M-ary orthogonal signals over independent and correlated fading channels. *Intl. J. Wireless Info. Net.*, 10(1):41–56, Jan. 2003.

[58] J. Cheng and T. Berger. Performance analysis for M-ary orthogonal FSK with hybrid selection/equal-gain combining over Nakagami fading channels. In *Proc. IEEE VTC*, volume 4, pages 2628–2632, Apr. 2003.

[59] B. Sklar. *Digital Communications: Fundamentals and Applications*. Prentice Hall, New Delhi, India, 2nd edition, Jan. 2001.

[60] L. Hanzo, S. X. Ng, T. Keller, and W. Webb. *Quadrature Amplitude Modulation: From Basics to Adaptive Trellis-Coded, Turbo-Equalised and Space-Time Coded OFDM, CDMA and MC-CDMA Systems*. Wiley-Blackwell, 2nd edition, 2004.

[61] C. N. Campopiano and B. G. Glazer. A coherent digital amplitude and phase modulation system. *IRE Trans. Commun.*, 10:90–95, Mar. 1962.

[62] C. J. Kim, Y. S. Kim, G. Y. Jung, and H. J. Lee. BER analysis of QAM with MRC space diversity in Rayleigh fading channel. In *Proc. IEEE PIMRC*, pages 482–485, Sep. 1995.

[63] C. J. Kim, Y. S. Kim, G. Y. Jeong, J. K. Mun, and H. J. Lee. SER analysis of QAM with space diversity in Rayleigh fading channels. *ETRI J.*, 17(4):25–35, Jan. 1996.

[64] J. Lu, T. T. Tjhung, and C. C. Chai. Error probability performance of L-branch diversity reception of MQAM in Rayleigh fading. *IEEE Trans. Commun.*, 46(2):179–181, Feb. 1998.

[65] A. Annamalai, C. Tellambura, and V. K. Bhargava. Exact evaluation of maximal-ratio and equal-gain diversity receivers for M-ary QAM on Nakagami fading channels. *IEEE Trans. Commun.*, 47(9):1335–1344, Sep. 1999.

[66] S. Seo, C. Lee, and S. Kang. Exact performance analysis of M-ary QAM with MRC diversity in Rician fading channels. *Electron. Lett.*, 40(8):485–486, Apr. 2004.

[67] I. A. Falujah and V. K. Prabhu. Performance analysis of MQAM with MRC over Nakagami-m fading channels. *Electron. Lett.*, 42(4):231–233, Feb. 2006.

[68] I. Falujah and V. K. Prabhu. Performance analysis of MQAM with MRC over Nakagami-m fading channels. In *Proc. IEEE WCNC*, volume 3, pages 1332–1337, Apr. 2006.

[69] G. K. Karagiannidis, N. C. Sagias, and D. A. Zogas. Error analysis of M-QAM with equal-gain diversity over generalised fading channels. *IEE Proc. Commun.*, 152(1):69–74, Feb. 2005.

[70] B. Sklar. Rayleigh fading channels in mobile digital communication systems, part I: Characterization. *IEEE Commun. Magz.*, 35(7):90–100, Jul. 1997.

[71] H. Bai and M. Atiquzzaman. Error modeling schemes for fading channels in wireless communications: A survey. *IEEE Commun. Surveys and Tutorials*, 5(2):2–9, 2003.

[72] J. D. Parsons. *The Mobile Radio Propagation Channel*. Prentech Press, London, 1994.

[73] M. K. Simon and M. -S. Alouini. *Digital Communication over Fading Channels*. John Wiley and Sons, NJ, USA, 2nd edition, Dec. 2004.

[74] M. Schwartz, W. R. Bennett, and S. Stein. *Communication Systems and Techniques*. McGraw-Hill, NY, USA, 1966.

[75] G. L. Stuber. *Principles of Mobile Communication*. Kluwer Academic/ Springer, New Delhi, India, 2nd edition, 2009.

[76] Y. Miyagaki, N. Morinaga, and T. Namekawa. Error probability characteristics for CPSK signal through m-distributed fading channel. *IEEE Trans. Commun.*, 26(1):88–100, Jan. 1978.

[77] U. Charash. Reception through Nakagami fading multipath channels with random delays. *IEEE Trans. Commun.*, 27(4):657–670, Apr. 1979.

[78] V. K. Garg. *Wireless Communications and Networking*. Morgan Kaufmann/ Elsevier, San Fransisco, CA, USA, Jun. 2007.

[79] R. G. Vaughan and J. B. Andersen. Antenna diversity in mobile communications. *IEEE Trans. Veh. Tech.*, 36(1):149–172, Nov. 1987.

[80] J. G. Andrews, A. Ghosh, and R. Muhamed. *Fundamentals of WiMAX: Understanding Broadband Wireless Networking*. Prentice Hall/ Pearson, NJ, USA, Feb. 2007.

[81] L. Zheng and D. N. C. Tse. Diversity and multiplexing: A fundamental tradeoff in multiple-antenna channels. *IEEE Trans. Info. Th.*, 49(5):1073–1096, May 2003.

[82] E. Lindskog. *Space-Time Processing and Equalization for Wireless Communications*. PhD thesis, Uppsala University, Uppsala, Sweden, Jun. 1999.

[83] M. Wennström. *On MIMO Systems and Adaptive Arrays for Wireless Communication: Analysis and Practical Issues*. PhD thesis, Uppsala University, Uppsala, Sweden, Sep. 2002.

[84] Y. Zhang, J. Cosmas, and Y. Song. Future transmitter/receiver diversity schemes in broadcast wireless networks. *IEEE Commun. Magz.*, 44(10):120–127, Oct. 2006.

[85] A. Alexiou and M. Haardt. Smart antenna technologies for future wireless systems: Trends and challenges. *IEEE Commun. Magz.*, 42(9):90–97, Sep. 2004.

[86] R. D. Murch and K. B. Letaief. Antenna systems for broadband wireless access. *IEEE Commun. Magz.*, 40(4):76–83, Apr. 2002.

[87] S. Serbetli and A. Yener. Transceiver optimization for multiuser MIMO systems. *IEEE Trans. Sig. Proc.*, 52(1):214–226, Jan. 2004.

[88] S. Sanayei and A. Nosratinia. Antenna selection in MIMO systems. *IEEE Commun. Magz.*, 42(10):68–73, Oct. 2004.

[89] J. Jootar, J. -F. Diouris, and J. R. Zeidler. Performance of polarization diversity in correlated Nakagami-m fading channels. *IEEE Trans. Veh. Tech.*, 55(1):128–136, Jan. 2006.

[90] W. C. Y. Lee. *Mobile Communications Engineering*. McGraw-Hill, 1998.

[91] P. Mattheijssen, M. H. A. J. Herben, G. Dolmans, and L. Leyten. Antenna-pattern diversity versus space diversity for use at handhelds. *IEEE Trans. Veh. Tech.*, 53(4):1035–1042, Jul. 2004.

[92] A. M. Sayeed and B. Aazhang. Joint multipath-Doppler diversity in mobile wireless communications. *IEEE Trans. Commun.*, 47(1):123–132, Jan. 1999.

[93] D. D. Hodges and R. J. Watson. An analysis of conditional site diversity: A study at Ka-band. *IEEE Trans. Antennas Propag.*, 57(3):721–727, Mar. 2009.

[94] K. S. Paulson, I. S. Usman, and R. J. Watson. A general route diversity model for convergent terrestrial microwave links. *Radio Science*, 41(3), 2006.

[95] J. Shin, K. Lee, A. Yener, and T. F. La Porta. On-demand diversity wireless relay networks. *Mobile Netw. Appl.*, 11(4):593–611, Aug. 2006.

[96] R. Janaswamy. Spatial diversity for wireless communications. In L. C. Godara, editor, *Handbook of Antennas in Wireless Communications*, chapter 19. CRC Press, 2002.

[97] D. Torrieri. *Principles of Spread-Spectrum Communication Systems*. Springer, 2005.

[98] S. Haykin and M. Moher. *Modern Wireless Communications*. Pearson Education, 2005.

[99] W. C. Jakes. A comparison of specific space diversity techniques for reduction of fast fading in UHF mobile radio systems. *IEEE Trans. Veh. Tech.*, 20(4):81–92, Nov. 1971.

[100] K. A. Norton. Transmission loss in radio propagation. *Proceedings of the IRE*, 41(1):146–152, Jan. 1953.

[101] A. S. Akki and F. Haber. A statistical model of mobile-to-mobile land communication channel. *IEEE Trans. Veh. Tech.*, 35(1):2–7, Feb. 1986.

[102] A. S. Akki. Statistical properties of mobile-to-mobile land communication channels. *IEEE Trans. Veh. Tech.*, 43(4):826–831, Nov. 1994.

[103] P. Almers, F. Tufvesson, and A. F. Molisch. Keyhole effect in MIMO wireless channels: measurements and theory. *IEEE Trans. Wireless Commun.*, 5(12):3596–3604, Dec. 2006.

[104] D. Gesbert, H. Bolcskei, D. A. Gore, and A. J. Paulraj. Outdoor MIMO wireless channels: Models and performance prediction. *IEEE Trans. Commun.*, 50(12):1926–1934, Dec. 2002.

[105] V. Erceg, S. J. Fortune, J. Ling, A. J. Rustako, and R. A. Valenzuela. Comparisons of a computer-based propagation prediction tool with experimental data collected in urban microcellular environments. *IEEE J. Select. Areas Commun.*, 15(4):677–684, May 1997.

[106] I. Z. Kovacs, P. C. F. Eggers, K. Olesen, and L. G. Petersen. Investigations of outdoor-to-indoor mobile-to-mobile radio communication channels. In *Proc. IEEE VTC*, volume 1, pages 430–434, Sep. 2002.

[107] J. Maurer, T. Fugen, K. Olesen, and W. Wiesbeck. Narrowband measurement and analysis of the intervehicle transmission channel at 5.2 GHz. In *Proc. IEEE VTC*, volume 3, pages 1274–1278, May 2002.

[108] G. Acosta, K. Tokuda, and M. A. Ingram. Measured joint Doppler delay power profiles for vehicle-to-vehicle communications at 2.4 GHz. In *Proc. IEEE GLOBECOMM*, volume 6, pages 3813–3817, Nov. 2004.

[109] A. G. Zajic and G. L. Stuber. Space-time correlated mobile-to-mobile channels: Modelling and simulation. *IEEE Trans. Veh. Tech.*, 57(2):715–726, Mar. 2008.

[110] C. S. Patel, G. L. Stuber, and T. G. Pratt. Simulation of Rayleigh-faded mobile-to-mobile communication channels. *IEEE Trans. Commun.*, 53(11):1876–1884, Nov. 2005.

[111] J. Salo, H. M. El-Sallabi, and P. Vainikainen. The distribution of the product of independent Rayleigh random variables. *IEEE Trans. Antennas Prop.*, 54(2):639–643, Sep. 2006.

[112] G. J. Byers and F. Takawira. Spatially and temporally correlated MIMO channels: Modeling and capacity analysis. *IEEE Trans. Veh. Tech.*, 53(3):634–643, May 2004.

[113] M. Pätzold, B. O. Hogstad, N. Youssef, and D. Kim. A MIMO mobile to mobile channel model: Part I-The reference model. In *Proc. IEEE PIMRC*, volume 1, pages 573–578, Sep. 2005.

[114] N. C. Sagias, G. K. Karagiannidis, and G. S. Tombras. Error-rate analysis of switched diversity receivers in Weibull fading. *Electron. Lett.*, 40(11):681–682, May 2004.

[115] H. -C. Yang and M. -S. Alouini. Performance analysis of multibranch switched diversity systems. *IEEE Trans. Commun.*, 51(5):782–794, May 2003.

[116] T. T. Tjhung, C. Loo, and N. P. Secord. BER performance of DQPSK in slow Rician fading. *Electron. Lett.*, 28(18):1763–1765, Aug. 1992.

[117] M. Tanda. Bit error rate of DQPSK signals in slow Nakagami fading. *Electron. Lett.*, 29(5):431–432, Mar. 1993.

[118] S. Chennakeshu and J. B. Anderson. Error rates for Rayleigh fading multichannel reception of MPSK signals. *IEEE Trans. Commun.*, 43:338–346, 1995.

[119] G. A. Baker and P. Graves-Morris. *Pade approximants.* Cambridge University Press, UK, 2nd edition, 1996.

[120] L. -C. Wang, W. -C. Liu, and Y. -H. Cheng. Statistical analysis of a mobile-to-mobile Rician fading channel model. *IEEE Trans. Veh. Tech.*, 58(1):32–38, Jan. 2009.

[121] B. Talha and M. Pätzold. Statistical modeling and analysis of mobile-to-mobile fading channels in cooperative networks under line-of-sight conditions. *Wireless Pers. Commun.*, 54(1):3–19, Jul. 2010.

[122] G. K. Karagiannidis, N. C. Sagias, and P. T. Mathiopoulos. $N*$Nakagami: A novel stochastic model for cascaded fading channels. *IEEE Trans. Commun.*, 55(8):1453–1458, Aug. 2007.

[123] H. Shin and J. H. Lee. Performance analysis of space-time block codes over keyhole Nakagami-m fading channels. *IEEE Trans. Veh. Tech.*, 53(2):351–362, Mar. 2004.

[124] K. Peppas and A. Maras. Performance evaluation of space-time block codes over keyhole Weibull fading channels. *Wireless Pers. Commun.*, 46:385–395, 2008.

[125] J. Salo, H. M. El-Sallabi, and P. Vainikainen. Impact of double-Rayleigh fading on system performance. In *Proc. IEEE ISWPC*, Jan. 2006.

[126] M. Uysal. Maximum achievable diversity order for cascaded Rayleigh fading channels. *Electron. Lett.*, 41(23):1289–1290, Nov. 2005.

[127] Y. Gong and K. B. Letaief. Space-time block codes in keyhole fading channels: Error rate analysis and performance results. In *Proc. IEEE VTC*, volume 4, pages 1903–1907, May 2006.

[128] M. Uysal. Diversity analysis of space-time coding in cascaded Rayleigh fading channels. *IEEE Commun. Lett.*, 10(3):165–167, Mar. 2006.

[129] N. H. Tran, H. H. Nguyen, and T. Le-Ngoc. Performance bounds of orthogonal space-time block codes over keyhole Nakagami-m channels. *IEEE Sig. Proc. Lett.*, 14(9):605–608, Sep. 2007.

[130] I. Z. Kovacs. *Radio Channel Characterisation for Private Mobile Radio Systems.* PhD thesis, Center for PersonKommunikation, Aalborg University, Denmark, Feb. 2003.

[131] D. Chizhik, G. J. Foschini, M. J. Gans, and R. A. Valenzuela. Keyholes, correlations, and capacities of multielement transmit and receive antennas. *IEEE Trans. Wireless Commun.*, 1(2):361–368, Apr. 2002.

[132] Y. L. Luke. *Integrals of Bessel Functions.* McGraw-Hill, NY, USA, 1962.

[133] A. P. Prudnikov, Y. A. Brychkov, and O. I. Marichev. *Integrals and Series: Special Functions*, volume 2. Gordon and Breach Science Publishers, Amsterdam, 1992.

[134] The Wolfram functions site. Hypergeometric Functions, HypergeometricU[a,b,z]–Specific values. [Online] Available: http://functions.wolfram.com/07.33.03.0446.01.

[135] S. -P. Yeh, S. Talwar, S. -C. Lee, and H. Kim. WiMAX femtocells: a perspective on network architecture, capacity, and coverage. *IEEE Commun. Magz.*, 46(10):58–65, Oct. 2008.

[136] H. T. Hui. The performance of the maximum ratio combining method in correlated Rician fading channels for antenna diversity signal combining. *IEEE Trans. Ant. and Prop.*, 53(3):958–964, Mar. 2005.

[137] S. W. Peters and R. W. Heath. The future of WiMAX: multihop relaying with IEEE 802.16j. *IEEE Commun. Magz.*, 47(1):104–111, Jan. 2009.

[138] F. Adachi. Error rate analysis of differentially encoded and detected 16APSK under Rician fading. *IEEE Trans. Veh. Tech.*, 45(1):1–11, Feb. 1996.

[139] J. Hagenauer, F. Dolainsky, E. Lutz, W. Papke, and R. Schweikert. The maritime satellite communication channel - Channel model, performance of modulation and coding. *IEEE J. Select. Areas Commun.*, 5(4):701–713, May 1987.

[140] D. Divsalar and M. K. Simon. Trellis coded modulation for 4800-9600 bits/s transmission over a fading mobile satellite channel. *IEEE J. Select. Areas Commun.*, 5(2):162–175, Feb. 1987.

[141] I. S. Barbounakis and A. M. Papadakis. Closed-form SER expressions for star MQAM in frequency non-selective Rician and Nakagami-*m* channels. *Int. J. Electron. and Commun. (AE)*, 59:417–420, 2005.

[142] T. S. Rappaport. Indoor radio communications for factories of the future. *IEEE Commun. Magz.*, 27:15–24, May 1989.

[143] R. J. C. Bultitude and G. K. Bedal. Propagation characteristics on microcellular urban mobile radio channels at 910 MHz. *IEEE J. Select. Areas Commun.*, 7(1):31–39, Jan. 1989.

[144] F. Wijk, A. Kegel, and R. Prasad. Assessment of a pico-cellular system using propagation measurements at 1.9 GHz for indoor wireless communications. *IEEE Trans. Veh. Tech.*, 44(1):155–162, Feb. 1995.

[145] E. Walker, H. -J. Zepernick, and T. Wysocki. Fading measurements at 2.4GHz for the indoor radio propagation channel. In *Proc. Int. Seminar on Broadband Communications: Accessing, Transmission, Networking*, pages 171–176, Feb. 1998.

[146] J. Medbo, H. Hallenberg, and J. -E. Berg. Propagation characteristics at 5 GHz in typical radio-LAN scenarios. In *Proc. IEEE VTC*, pages 185–189, 1999.

[147] W. C. Lindsey. Error probabilities for Rician fading multichannel reception of binary and N-ary signals. *IEEE Trans. Info. Th.*, 10:339–350, 1964.

[148] J. A. Roberts and J. M. Bargallo. DPSK performance for indoor wireless Rician fading channels. *IEEE Trans. Commun.*, 42(2/3/4):592–596, Feb./Mar./Apr. 1994.

[149] J. Sun, I. S. Reed, L. Technol, and N. J. Whippany. Linear diversity analyses for M-PSK in Rician fading channels. *IEEE Trans. Commun.*, 51(11):1749–1753, Nov. 2003.

[150] M. -S. Alouini and M. K. Simon. Generic form for average error probability of binary signals over fading channels. *Electron. Lett.*, 34(10):949–950, May 1998.

[151] S. Haghani and N. C. Beaulieu. Symbol error rate performance of M-ary NCFSK with S + N selection combining in Rician fading. In *Proc. IEEE ICC*, volume 4, pages 2500–2505, May 2005.

[152] S. Haghani and N. C. Beaulieu. M-ary NCFSK with S + N selection combining in Rician fading. *IEEE Trans. Commun.*, 54(3):491–498, Mar. 2006.

[153] M. S. Patterh, T. S. Kamal, and B. S. Sohi. Performance of coherent square MQAM with L^{th} order diversity in Rician fading environment. In *Proc. IEEE VTC*, volume 1, pages 141–143, 2001.

[154] C. Yang, G. Bi, and A. R. Leyman. MRC receiver performance with MQAM in correlated Rician fading channels. In *Proc. IEEE/SP Workshop on Stat. Sig. Procs.*, pages 210–212, Aug. 2001.

[155] A. A. Ali and K. Alkhudairi. BER for M-QAM with space diversity and MRC in Rician fading channels. In *Proc. IEEE MCWC*, pages 67–72, Sep. 2006.

[156] A. J. Goldsmith and S. -G. Chua. Variable-rate variable-power MQAM for fading channels. *IEEE Trans. Commun.*, 45(10):1218–1230, Oct. 1997.

[157] H. Zhang and T. A. Gulliver. Error probability for maximum ratio combining multichannel reception of M-ary coherent systems over flat Ricean fading channels. In *Proc. IEEE WCNC*, volume 1, pages 306–310, Mar. 2004.

[158] L. Najafizadeh and C. Tellambura. BER analysis of arbitrary QAM for MRC diversity with imperfect channel estimation in generalized Ricean fading channels. *IEEE Trans. Veh. Tech.*, 55(4):1239–1248, Jul. 2006.

[159] Y. Ma, R. Schober, and D. Zhang. Exact BER for M-QAM with MRC and imperfect channel estimation in Rician fading channels. *IEEE Trans. Wireless Commun.*, 6(3):926–936, Mar. 2007.

[160] A. Annamalai. Analysis of selection diversity on Nakagami fading channels. *Electron. Lett.*, 33(7):548–549, Mar. 1997.

[161] K. Noga. The performance of binary transmission in slow Nakagami-fading channels with MRC diversity. *IEEE Trans. Commun.*, 46(7):863–865, Jul. 1998.

[162] V. Aalo and S. Pattaramalai. Average error rate for coherent MPSK signals in Nakagami fading channels. *Electron. Lett.*, 32(17):1538–1539, Aug. 1996.

[163] S. J. Baik, S. Y. Choi, D. W. Yoon, K. J. Lee, and Y. T. Han. Analysis of selection diversity for MPSK signals in Nakagami fading channels. In *Proc. IEEE ICPWC*, pages 138–141, Dec. 1997.

[164] A. Annamalai, C. Tellambura, and V. K. Bhargava. Unified analysis of MPSK and MDPSK with diversity reception in different fading environments. *Electron. Lett.*, 34(16):1564–1565, Aug. 1998.

[165] G. Fedele. Switched dual diversity reception of M-ary DPSK signals on Nakagami fading channels. *Wireless Pers. Commun.*, 8(1):53–71, Aug. 1998.

[166] C. C. Chai and T. T. Tjhung. Error probabilities for MRC diversity reception of MDPSK in slow Nakagami fading. *Wireless Pers. Commun.*, 16(2):149–172, Feb. 2001.

[167] N. Ekanayake, M. B. Dissanayake, and D. A. D. C. Dhanapala. MRC diversity reception of M-ary DPSK signals in slow Nakagami fading channels. *Electron. Lett.*, 42(4):230–231, Feb. 2006.

[168] H. -C. Yang, M. -S. Alouini, and M. K. Simon. Average error rate of NCFSK with multi-branch post-detection switched diversity. *Wirel. Commun. Mob. Comput.*, 4(4):351–367, Jun. 2004.

[169] R. M. Radaydeh and M. M. Matalgah. Closed-form formula for the average BER of noncoherent multi-level orthogonal signals over nonidentical Nakagami fading channels. In *Proc. ACM IWCMC*, pages 905–910, Jul. 2006.

[170] M. S. Patterh, T. S. Kamal, and B. S. Sohi. BER performance of MQAM with L-branch MRC diversity reception over correlated Nakagami-m fading channels. *Wirel. Commun. Mob. Comput.*, 3(3):397–406, May 2003.

[171] D. Yoon, D. -I. Chang, N. -S. Kim, and H. -S. Woo. Linear diversity analysis for M-ary square quadrature amplitude modulation over Nakagami fading channels. *ETRI J.*, 25(4):231–2375, Aug. 2003.

[172] A. Annamalai, C. Tellambura, and V. K. Bhargava. A general method for calculating error probabilities over fading channels. *IEEE Trans. Commun.*, 53(5):841–852, May 2005.

[173] G. Fedele. Error probability for diversity detection of binary signals over Nakagami fading channels. In *Proc. IEEE PIMRC*, volume 2, pages 607–611, Sep. 1994.

[174] M. Blanco. Diversity receiver performance in Nakagami fading. In *Proc. IEEE Southeastcon*, pages 529–532, 1983.

[175] S. Okui. Probability of co-channel interference for selection diversity reception in the Nakagami m-fading channel. *IEE Proc. Commun.*, 139(1):91–94, Feb 1992.

[176] R. Sannegowda and V. A. Aalo. Performance of selection diversity systems in a Nakagami fading environment. In *Proc. IEEE Southeastcon*, pages 190–195, Apr. 1994.

[177] O. C. Ugweje and V. A. Aalo. Performance of selection diversity system in correlated Nakagami fading. In *Proc. IEEE VTC*, volume 3, pages 1488–1492, May 1997.

[178] M. Lo and W. H. Lam. Probability of symbol error for MPSK, MDPSK and noncoherent MPSK with MRC and SC space diversity in Nakagami-m fading channel. In *Proc. IEEE WCNC*, volume 3, pages 1427–1431, Sep. 2000.

[179] M. Lo and W. H. Lam. Performance analysis of bandwidth efficient coherent modulation schemes with L-fold MRC and SC in Nakagami-m fading channels. In *Proc. IEEE PIMRC*, volume 1, pages 572–576, Sep. 2000.

[180] O. C. Ugweje. Selection diversity for wireless communications in Nakagami-fading with arbitrary parameters. *IEEE Trans. Veh. Tech.*, 50(6):1437–1448, Nov. 2001.

[181] R. Annavajjala, A. Chockalingam, and L. B. Milstein. Performance analysis of coded communication systems on Nakagami fading channels with selection combining diversity. *IEEE Trans. Commun.*, 52(7):1214–1220, Jul. 2004.

[182] G. Fedele. Error probability for detection of M-DPSK signals in slow nonselective Nakagami fading. *Electron. Lett.*, 30(8):620–622, Apr. 1994.

[183] G. Fedele, L. Izzo, and M. Tanda. Dual diversity reception of M-ary DPSK signals over Nakagami fading channels. In *Proc. IEEE PIMRC*, volume 3, pages 1195–1201, Sep. 1995.

[184] G. Fedele. N-branch diversity reception of M-ary DPSK signals in slow and nonselective Nakagami fading. *Eur. Trans. Telecommun.*, 7(2):119–123, Mar.-Apr. 1996.

[185] C. M. Lo and W. H. Lam. Average SER for M-ary modulation systems with space diversity over independent and correlated Nakagami-m fading channels. *IEE Proc. Commun.*, 148(6):377–384, Dec. 2001.

[186] L. C. Choo and T. T. Tjhung. BER performance of DQPSK in Nakagami fading with selection diversity and maximal-ratio combining. *IEEE Trans. Commun.*, 48(10):1618–1621, Oct. 2000.

[187] Q. T. Zhang and H. G. Lu. A general analytical approach to multi-branch selection combining over various spatially correlated fading channels. *IEEE Trans. Commun.*, 50(7):1066–1073, Jul. 2002.

[188] Y. Chen and C. Tellambura. Distribution functions of selection combiner output in equally correlated Rayleigh, Rician, and Nakagami-m fading channels. *IEEE Trans. Commun.*, 52(11):1948–1956, Nov. 2004.

[189] J. Reig, L. Rubio, and V. M. R. Penarrocha. Performance of dual selection combiners over correlated Nakagami-m fading with different fading parameters. *IEEE Trans. Commun.*, 54(9):1527–1532, Sep. 2006.

[190] A. -H. S. Al-Hussien and M. M. Banat. Probability of error analysis of predetection generalized selection combining receivers with correlated unbalanced Nakagami branches. *Wireless Commun. Mobile Comput.*, 7(6):689–701, Aug. 2007.

[191] N. Kong and L. B. Milstein. Asymptotic symbol error rate for selection combining on Nakagami-m fading channels. In *Proc. IEEE GLOBECOM*, pages 1–5, Nov.-Dec. 2008.

[192] M. -S. Alouini and M. K. Simon. Performance of coherent receivers with hybrid SC/MRC over Nakagami-m fading channels. *IEEE Trans. Veh. Tech.*, 48(4):1155–1164, Jul. 1999.

[193] R. M. Radaydeh. Average error performance of M-ary modulation schemes in Nakagami-q (Hoyt) fading channels. *IEEE Commun. Lett.*, 11(3):255–257, Mar. 2007.

[194] M. K. Simon and M. -S. Alouini. Some new results for integrals involving the generalized Marcum Q function and their application to performance evaluation over fading channels. *IEEE Trans. Wireless Commun.*, 2(4):611–615, Jul. 2003.

[195] M. K. Simon. *Probability Distributions Involving Gaussian Random Variables: A Handbook for Engineers and Scientists*. Kluwer Academic, Norwell, MA, USA, 1st edition, 2002.

[196] J. F. Paris. Nakagami-q (Hoyt) distribution function with applications. *Electron. Lett.*, 45(4):210–2112, Feb. 2009.

[197] R. M. Radaydeh and M. M. Matalgah. Average ber analysis for M-ary FSK signals in Nakagami-q (Hoyt) fading with noncoherent diversity combining. *IEEE Trans. Veh. Tech.*, 57(4):2257–2267, Jul. 2008.

[198] G. Fraidenraich, J. C. S. S. Filho, and M. D. Yacoub. Second-order statistics of maximal-ratio and equal-gain combining in Hoyt fading. *IEEE Commun. Lett.*, 9(1):19–21, Jan. 2005.

[199] D. Krstić, P. Nikolić, G. Stamenović, and D. Stefanović. Performance analysis of generalized selection combiner in the presence of Hoyt fading. In *Proc. IEEE TELSIKS*, pages 599–602, Oct. 2009.

[200] Y. -C. Ko, M. -S. Alouini, and M. K. Simon. Analysis and optimization of switched diversity systems. *IEEE Trans. Veh. Tech.*, 49(5):1813–1831, Sep. 2000.

[201] H. A. Taha. *Operations Research: An Introduction*. Prentice Hall of India, New Delhi, India, 6th edition, 1999.

[202] N. C. Beaulieu and C. Cheng. Efficient Nakagami-m fading channel simulation. *IEEE Trans. Veh. Tech.*, 54(2):413–424, Mar. 2005.

[203] G. Marsaglia and W. W. Tsang. A simple method for generating gamma variables. *ACM Trans. Math. Software*, 26(3):363–372, Sep. 2000.

[204] N. Youssef, C. -X. Wang, and M. Pätzold. A study on the second order statistics of Nakagami-Hoyt mobile fading channels. *IEEE Trans. Veh. Tech.*, 54(4):1259–1265, Jul. 2005.

[205] R. S. Hoyt. Probability functions for the modulus and angle of the normal complex variate. *Bell Syst. Tech. J.*, 26:318–359, Apr. 1947.

Lightning Source UK Ltd.
Milton Keynes UK
UKHW010648190421
382245UK00001B/44